MAN IS
THE PREY

The Lion

MAN IS
THE PREY

James Clarke

Stein and Day / *Publishers* / New York

Copyright © 1969 by James Clarke
Library of Congress Catalog Card No. 70-87953
All rights reserved
Printed in the United States of America
Designed by David Miller
Stein and Day/*Publishers*/7 East 48 Street, New York, N.Y. 10017
SBN: 8128-1250-6

This book is dedicated to George Gray, who died in Nairobi Hospital after being mauled by a lion; to Hemidi Ngoe, who was eaten by a lion in Tanganyika; to the 434 victims of the Champawat tigress; to the 185 men, women, and children killed by the leopard of Rudraprayag; to old Bill Judd, who was killed by a wounded elephant in East Africa; to the child victims of the man-eating hyenas of Malanje; to Sam Adams, eaten by a grizzly bear in Montana; to Andries Johannes Steyn, killed by a hippopotamus on the Okavango; to Charles Gray, brother of George Gray, killed by a buffalo; to the uncounted thousands killed by sharks; to the 40,000 who die each year from snake bite; to the hundreds of Africans eaten by crocodiles annually; to Dirk Avontuur, killed by an ostrich in Karoo; to former Mayor A. Barnard, killed by a bee in Kempton Park; to seven-year-old Dominic Taylor, killed by a cougar in British Columbia; and to the countless other victims of man-killers and man-eaters.

Contents

CONTENTS

ACKNOWLEDGMENTS

A great deal of the information in this book is drawn from the works of many writers on animals, and among these I would especially like to acknowledge Richard Perry, R. L. Ditmars, George G. Rushby, John Pollard, Jim Corbett, Kenneth Anderson, W. Robert Foran, and C. J. P. Ionides. Other important sources are mentioned in the bibliography.

I also owe a deep debt of gratitude to the World Health Organization's Brazzaville office, the *Star* in Johannesburg, whose files have been invaluable, the Toronto *Star* for several snippets of information which were always promptly given, and to the Wildlife Protection and Conservation Society of South Africa.

The photographs in this book are reproduced by kind permission of John Pitts, the *Star* (Johannesburg), the *Daily News* (Durban), the *Evening News* (Port Elizabeth), P. D. Hugo, and T. E. Reilly.

In any book the author naturally tends to embroil his close friends and colleagues. In this regard I can never really adequately thank such people as John Pitts, who went to endless trouble and took not a few risks in obtaining animated pictures of the deadly Indian cobra, the puff adder, the lion, leopard, cheetah, and tiger, the white rhinoceros, the African elephant, and many other animals—almost all of them in the wilds.

I received invaluable assistance from Rod Patterson and Bernie J. Keyter (herpetologists), Ian Player (Nature Conservation of Zululand), A. H. (Lammy) Lambrechts (Johannesburg Zoo), Clive Cowley (an expert on Bushmen), Patrick Turner (who, unasked, dug out many of the quotations which head the chapters), Rod Barrett, M. Johnston, S. Payne, K. Cunningham, and Mrs. Yvonne van der Walt.

I would also like to acknowledge the help received in so many diverse ways from Mrs. A. E. Clarke, Mr. Grgo Babaya, and, finally, from my wife, Lenka, who suffered in a way that only the wives of writers can appreciate.

I
MAN-KILLERS

The African Buffalo

Man-Killers

"Nature does not care whether the hunter
slays the beast or the beast the hunter. She
will make good compost of them both and
her ends are prospered whichever suc-
ceeds."

John Burroughs

This book is written in the belief that people like to know
precisely what is eating them. It records briefly some of the
epic and stirring duels between man and his natural enemies.
It also attempts to put them in their biological perspective.

I would be disappointed if it were taken to be an indict-
ment of the world of wild animals. It is not that at all. Its
primary purpose is to investigate the actions and motives of
man-killers and man-eaters and to set the record straight regarding some species that have been maligned by fiction writers
or through lack of understanding.

Advanced and sophisticated though man is, he is still a
vulnerable animal. Without a gun he is no match for predators
of even the meanest intelligence, for he possesses neither fangs
nor claws, tough hide nor fleetness of foot. It might have been
this inequality that drove him to become so obsessed with
weapons that today he has emerged as the second greatest destroyer of human life in the animal kingdom.

Every continent has its share of dangerous wild animals
but, obviously, there are some places where the danger is
greater than in others. Possibly the most dangerous area on
earth is Central Africa. In this uncomfortable region mosquitoes
and tsetse flies kill thousands of people annually. Scorpions, bees,
black widow spiders, and a score of different kinds of snakes
bite or sting hundreds of Central Africans to death. Lions,

leopards, and hyenas take a steady human toll too, and so do elephants, rhinoceroses, hippopotamuses, and buffaloes. In the rivers glide the crocodiles, which are possibly the worst man-eaters of all, and along the coast cruise sharks and other marine killers.

In terms of people killed, Central Africa might have a rival in a part of Burma, where V. F. M. FitzSimons claims 30 people in every 1000 die of snake bite and in addition thousands are killed by insect-borne diseases. Other parts of Asia are not much better off and have to contend with such creatures as the mugger crocodile, some of the world's most dangerous snakes, elephants, rhinoceroses, tigers, leopards, and bears, and, in the north, wolves.

The Americans have a fair number of man-killers including a score of lethal snakes, crocodiles, the world's only venomous lizards, bears, wolves, cougars, jaguars, black widow spiders, disease-carrying insects and fleas, and certain aquatic and sea animals.

Australasia has some of the most voracious sharks in the world and a few coastal killers such as sea wasps and cone shells. Australia itself is unique in that there are more venomous snakes than harmless ones. But New Zealand, Australia's sister country, is one of the safest places on earth, having no snakes and, in fact, no man-killing animals except a venomous spider and coastal sharks.

The least perilous continent is Europe: its wolves and bears claim very few victims nowadays and it has a notable lack of highly venomous snakes. Possibly the safest large country in the world is Ireland, which has nothing more lethal than the bumble bee.

The most persistent man-killing creatures in the world are insects. Man's struggle with the insect world has still not been conclusively won, although it could have been won decades ago if we had spent as much on combatting the insect plagues as we have spent on combatting ourselves. Man takes second place only to the insect world in the number of men he kills:

then come the snakes. The World Health Organization (WHO) estimated in 1954 that 40,000 people are killed annually by snakes. Of these, stated the WHO report, 70 percent died in Asia. The figure of 40,000 was described by WHO as "conservative." The figure may, in fact, be far short of the mark. Some authorities believe the figure should be nearer 80,000, and V. F. M. FitzSimons quotes "an authoritative source" giving the annual death toll in India alone as 30,000. The trouble with snake-bite statistics is that it is impossible to keep a tally in areas where written records do not exist, and it is these remote, undeveloped areas that often have a high number of snake bites.

Although snakes are undoubtedly the biggest man-killers among wild animals, the sharks and the crocodiles are contenders for the title of biggest man-eater. In their natural element neither show undue fear of man, and in many parts of their range they treat man as a welcome addition to their diet. Most other man-killing or man-eating creatures are members of species that are normally unaggressive toward man; therefore when they start to hunt down human beings it must be taken as abnormal behavior.

As a man-eater the tiger is the busiest among the big cats and has eaten possibly as many as 2000 people a year in fairly recent times. The lion is second among the big cats and is unusual in that it will often develop a craving for human flesh and will then eat nothing else. Such a lion might require fifty human victims a year in order to live. The leopard, on the rare occasions it turns man-eater, can do tremendous damage, and individuals have established some extraordinary records. However, wounded leopards, notoriously vindictive though they are, are more often than not successfully beaten off when they ambush unarmed men. The hyena is another serious problem as a man-eater in parts of Africa, and the elephant, as a man-killer, exacts a fairly heavy toll. There have been delinquent bears from time to time, and the wolf in Asia is still a serious menace.

George Bernard Shaw had a theory that meat-eating animals such as lions and tigers (and, he hints, humans) tend to skulk when danger approaches, while vegetarians such as buffaloes and rhinoceroses (and no doubt vegetarian Shaw himself) are bold and aggressive. It is worth noting that vegetarian animals are more often than not easier to fool and to avoid. The carnivores are usually the more sly ("skulking" is a poor adjective) and once they turn rogue they are definitely the more difficult to hunt. The fact that carnivores tend to "skulk" rather than charge as a means of defending themselves seems to indicate intelligence rather than stupidity.

In Africa hunters refer to the five most dangerous big-game animals—the elephant, buffalo, rhinoceros, lion, and leopard—as the "big five." International hunters talk of the "big seven," which include the above five, plus the grizzly and the tiger. Which of these is the most dangerous to the hunter? There is a tremendous divergence of opinions: George Rushby, a hunter who shot hundreds of elephants, found that lions were dull stuff in comparison. Hunters of equal renown such as von Schellendorff, Dennis D. Lyell, E. Tempoton-Perkins, and John F. Burger regarded lions as the most dangerous. Robert Foran, who hunted all seven, considered the tiger and the African elephant equally dangerous and placed them above the lion. C. P. Ionides, who did a tremendous amount of shooting in Africa, put the leopard on top, as do a number of others. The latter authorities say that if you mess up your first shot on a leopard you rarely get a chance for a second. But ignoring what all of them say and going purely by the number of hunters who die, and by what killed them, the African elephant emerges as being the most dangerous. Also, from what I know of hunters and their experiences, they feel more qualms when setting out after a wounded or rogue elephant than they feel going after any other quarry.

Richard Perry poses an intriguing question in his *The World of the Tiger:* Why does the tiger not treat man as natural food? The question could as properly be asked of the lion and other large meat-eaters such as leopards, jaguars, wolves, and bears.

In the case of the tiger, Perry reasons, "aboriginal man provides a tiger with as easy a kill as a small mousedeer, so that if human prey were natural to tigers it is not unreasonable to suggest that all jungle tribes would long since have been exterminated. . . . The possible answer would seem that the tigers have been established in their jungles, and in their pattern of killing, for many thousands of years before they first came into contact with man whose unjungle-like appearance . . . with his upright posture and swinging arms and legs, rendered him an object to be curious of and suspicious of, rather than potentially edible, until tigers had had considerable experience of his ways." There are other good reasons for the carnivores' natural reluctance to eat men as a matter of course: lions, tigers, and leopards invariably go for one of the more prolific species of animal as these are naturally easier to come across. In few areas hunted by big cats is man a particularly common species.

L. S. B. Leakey claimed that man has survived in the face of competition from wild animals only because he smells. "Nature endowed us with something of either a nasty taste or smell where the carnivores are concerned," argued Leakey at a press conference in New York in March 1967. But it seems much more likely that man has *become* unpalatable *since* his dawn days, when he rapidly emerged as a crafty hunter whose hunting methods and aggressiveness must have gradually earned him the fear of the carnivore. At Makapansgat in the northern Transvaal there is evidence that the "southern ape-man," *Australopithecus prometheus,* knew fire two million years ago, in which case he almost certainly would have used it to frighten animals from their caves or from their kill. Since man first flung a stick or a stone at an animal, his occupation—better still, his preoccupation—has been weapons and hunting. He had to learn to be a superefficient hunter or perish, and by the time the Stone Age was in full swing not even the biggest creatures on earth were safe from the incredible little hunter. Early man in Europe as well as the prehistoric American Indians killed mastodons and mammoths fifteen thousand years ago—probably by ringing them with fire and driving them over cliffs. Thus it was that carnivores and other dangerous animals learned to slink away when they smelled the scent of man. Today, in most parts of the wilds, the faintest whiff of man's odor automatically triggers off a fear response in the mightiest creatures.

There are indications, though, that once a carnivore has killed its first man (quite often accidentally such as when a herdboy gets in the way while it is attacking cattle or sheep) it finds the taste agreeable. Having once tasted man it will come back for more. This addiction to human flesh occurs in lions, and is also found, sometimes to a lesser extent, in tigers, leopards, and sharks.

Although early man probably used fire with sometimes devastating effect, the use of an ordinary campfire to keep away wild animals is highly overrated. In the wilderness of America and the bush of Africa, campfires are usually not only completely useless in keeping the really dangerous animals away but sometimes even attract them. In Africa the hippopotamus is notorious for charging campfires, and so is the black rhinoceros. In the United States grizzly bears have deliberately charged campfires and mauled campers. There are many instances of man-eating lions and tigers snatching up victims sleeping in the "security" of a fire and there are scores, if not hundreds, of cases where hyenas have snapped off the faces of Africans sleeping beside fires.

Big-game hunter John Taylor, who had a great deal of experience in the man-eater country in Central Africa, found fires useless but swore that his three-hundred-candlepower storm lantern was effective; "even the most determined man-eater would never venture into the dazzling blaze of light." When Taylor talks of "determined man-eaters" he is selecting his words carefully: the infamous man-eaters of Tsavo—that pair of bold lions which killed an unknown number of people at the turn of the century—had a complete disregard for fire. Almost nothing would stop these lions once they had selected their victim, not even firebrands hurled at them. In fact, they were so disdainful of fire that they ate some of their victims in the light of it.

If there is one thing that will put many dangerous animals off guard and even keep them at bay it is the human voice. In a magnificently pedantic work called *Wood's Illustrated Natural History* (1854) by Reverend J. G. Wood, "a certain Mr. Cummings," caught by a wounded lion, said to it, "Take it easy!" The lion dropped him and fled. It was not so much what Mr. Cummings said, of course; it was just the sound of his voice. Richard Perry says, "Man's noise was particularly upsetting [to the tiger] and noise is the one deterrent that will keep the most confirmed man-eaters at bay." Africans resort to shouting to keep hippopotamuses, elephants, and wild pigs off their crops at night, and a sudden shout has often been enough to confuse a charging rhinoceros. South African farmer R. de la B. Barker, writing in *African Wild Life,* told how one of his African workmen turned a buffalo nine times in succession by screaming at it. I have seen baboons using the same device when attacked by a leopard; the "elders" advanced screaming, not in terror but in order to intimidate. Swimmers have shouted underwater, successfully scaring off sharks. I was talking about the effect of the human voice on lions to Dick Chipperfield, Sr., of the famous circus-owning family, a few years ago, and he told me that if a lion "cut up rough" in the ring or threatened to "clobber" him he would shout. He said, "It usually puts them off guard. Then you can quickly assert yourself again."

The reactions of a human community to an attack by an animal upon one of its members are interesting. An example of the outrage that sweeps a community was demonstrated in the summer of 1957–58 when five bathers were killed by sharks off the Natal coast. The Navy was called in to depth-charge the area, marine-research organizations were galvanized into action, money poured in from a shocked public for research work, and there was talk of instituting air patrols to spot and bomb sharks. Newspapers carried banner headlines of the attacks, and many hotels along the seaside had to close down as guests canceled their vacation plans. It did not make much sense to the lonely, forgotten, short-of-funds road-safety organizers; they were busy totaling the usual couple of dozen people who had been killed on the road as they made their way to these resorts.

Why is it that when a man-eating animal takes a man almost the entire population becomes incensed? Is it some latent fear dating back to precivilization that makes us so hysterical? Are we perhaps like the monkeys and baboons, who rush to a spot where one of their numbers has been snatched by a leopard and shriek and scream in unison? Or is it something more simple—like indignation? Charles A. Guggisberg in his brilliant book *Simba* probably hits near the mark:

"It is easily understandable that human imagination is always stirred by stories of man-eaters, be they big cats, wolves, crocodiles or sharks. One experiences a by no means disagreeable shudder—provided there is no chance of being chosen as the next meal—as well as a feeling of outraged amazement that lowly animals can have the effrontery to gobble up the 'lord of creation.' From this feeling probably stems the strange effort to show the man-eating habits, especially of the big cats, as something absolutely out of the ordinary—something almost monstrous. Thus a very eminent scientist compares the man-eating lion with a homicidal maniac. I admit that I am unable to follow his reasoning. . . . a man-eater does not harm individuals of his own species. He kills other animals, only those victims are for once not zebras or wildebeeste, they belong to that strange species known as *Homo sapiens*. Why should this be so monstrous?"

Shaw put it more bluntly when he observed that if a man wants to murder a tiger he calls it sport but "when a tiger wants to murder him he calls it ferocity."

Just how dangerous are the wilds? Anybody who has grown used to living in the African bush, the Indian jungle, the towering Rocky Mountains, or the snowy wastes of the Arctic will tell you he feels a great deal safer there than in the cities. The wilds, generally speaking, are safe enough whether they are wolf-infested, lion-infested, or haunted by tigers. But things can go wrong and do. João Augusto Silva, a hunter of great renown in Portuguese East Africa, states: "No animal will attack [a man] except in abnormal circumstances. Apart from these uncommon occasions an attack is as rare as one man assaulting another for no rhyme or reason. Occurrences of this kind should be considered as pathological abnormalities. All wild animals are timid, sensitive and suspicious." He is perhaps overgeneralizing, but it is accepted by behaviorists that most animals have set "flight distances"—the distances at which they retreat when they feel a man has approached near enough. F. J. Pootman tried to measure these distances and came to the conclusion that lions retreat when a man gets to within 80 yards, elephants at 150 yards (in open country), crocodiles at 150 yards (when discovered out of water), hyenas at 21 feet, and lizards at 3 feet. Most antelopes rush off at about 200 yards—this would probably go for the buffaloes too. There are some animals a person may collide with in certain types of country. Men have walked into dozing elephants and rhinos, and on one occasion I saw a hunter run smack into a lioness. She got a bigger fright than he and bolted like a startled rabbit.

Generally, most animals would rather flee than face man. The wilds are, under normal circumstances, safer than built-up areas, where there are so many ways of losing one's life and where urban man is often a dangerous predator.

Men have been killed and eaten in thousands of freak incidents involving domestic and wild animals. Many authenticated stories refer to pigs eating people, especially very young children. Cows, bulls, horses, and dogs frequently kill people. In 1960, according to Bernard Grzimek, 169 Germans were killed by horses, cattle, and pigs and a further 36,549 were injured by domestic animals. Cats too have taken their toll, especially in the suffocation of newborn children by lying on their faces, possibly having smelled milk on them. In September

1965, a coroner in Perth, Australia, returned a verdict of death by accident after finding that a child had been suffocated by a cat; it was the second such verdict in two years. Pet dogs (quite apart from rabid animals) take a regular toll of human life and the incidence is not confined to any one breed. Often it is an old family pet which suddenly loses its reason but more often than not the killers are dogs which have been encouraged to be savage and have been kept either on a chain or in a confined area.

The camel has bitten several handlers to death. A camel in Texas trampled its Army handler to death after he had overloaded it. In the East, camel drivers say that camels can work up an extreme hatred for a man who overloads them and will often strike because of this. To get the animal back into a tractable frame of mind the driver will dump his clothes at its feet and the camel will then bite and trample them. Afterward the animal is usually well behaved. The bite of a camel is as powerful as the bite of a zebra; both are capable of severing a man's limb.

The giraffe, normally a placid, timid beast, can occasionally turn rogue but it is rarely a man-killer and therefore is not included in this book. The only death I have come across that was certainly caused by a giraffe was on the Mombasa–Nairobi road and involved a motorist. A giraffe, possibly injured in a fight, ran up and put its giant forefeet through the windshield, killing the driver instantly.

Once one realizes the rate at which animals are killing twentieth-century man in spite of our awesome array of weapons, one can get an inkling of what it must have been like before we were properly armed against our natural enemies—before we had guns to shoot man-eaters or medicines to fight the diseases brought by insects. No doubt early man faced the most severe competition in the animal world. This competition, from insects, reptiles, mammals, birds and fishes, must at times have very nearly overwhelmed the human race. Only our astounding capacity for reproduction could have saved us from annihilation.

II
GIANTS THAT KILL

The African Elephant

The Elephant

"The Elephant is a gentleman."

Rudyard Kipling

The elephant was nature's last fling at making monsters and it has proved a fairly successful one. In elephant country, man is never really welcome.

The forests of Central Africa are quiet, awesome places. There is no breeze, and when a bird sings out its voice is quickly absorbed by the hush and its notes fall flat and dead. When you walk in these primeval forests you tread silently, and if it becomes necessary to pass a remark you do so half in a whisper. I don't know why. Maybe it is because you feel intimidated by the great trees that rise into the dappled green light of the canopy leaves, dwarfed by plants and disquieted by a sensation of being watched. Man left the forests when he departed from the way of the apes millions of years ago and he no longer feels at ease there.

There are plenty of paths to follow, even where man has never trod: paths made by generations of buffaloes and rhinoceroses and fleeting, wraithlike antelopes. When you follow these paths past the moss-covered tree trunks and the cold, dripping monkey vines you are aware of one thing: you are trespassing in elephant country.

Zoka Forest, Uganda, is such a place as this: quiet and menacing. It draws its nourishment from the rich, damp earth that drains into the young Nile at Gondokoro. The Zoka has been the scene of many elephant hunts, some successful, a few disastrous. Among the most classic in modern times was one described by W. Robert Foran.

Foran, one of this century's great hunters and one of the last of the "old school," was carrying a .256 Mannlicher-Schoenauer. Most hunters would consider the rifle too light for elephant but Foran at that time considered it the best all-around rifle. On his heels came Hamisi bin Baraka, carrying the heavier .350 Rigby-Mauser. Hamisi, one of the last of the old-school gunbearers, was like his master: a man who never flinched. He had been with Foran for a number of years.

The two men were deep into the Zoka Forest following a game trail and keeping a sharp lookout for buffalo. Foran had been walking, head down, examining the spoor when he had cause to look up—straight into the smoldering eyes of a massive bull elephant, whose great tusks gleamed in the half-light. The elephant immediately charged, its heavy trunk dangling. It is a certain indication that an elephant means business when it comes forward in that fast, shuffling trot with its trunk hanging and the tip turned inward. Foran knew it was useless to retreat. He was facing the most dangerous target in the world, a charging rogue elephant.

A shot to the brain in such circumstances is terribly risky, for the critical area is a mere six inches wide and the enormous, one-ton head is usually moving from side to side. A heart shot is also tricky because the hanging trunk can deflect or help absorb even the heaviest bullets. The bull was within fifty paces before Foran was able to bring his rifle to his shoulder. Already the elephant was reaching forward with its trunk, to seize him. At the same time it began a series of unearthly screams. Foran took advantage of the groping trunk and aimed beneath it. The first bullet crashed into the chest cavity, bringing the monster momentarily to its knees. But a .256 has little stopping power when used against an elephant. Before the elephant could rise, Foran planted a second bullet between its eyes. The bull, screaming with rage, stood up and continued to charge. So fast was the animal's advance that Foran had no time to grab the heavier rifle from Hamisi, who, for some reason, had not opened fire. At ten yards Foran fired blindly but the bullet spent itself uselessly somewhere in the seven tons of moving gray hulk. A second later the elephant's trunk curled about Foran's waist, swung him clear of the ground and hurled him into the air. The hunter landed in a thorn tree; bruised, almost shirtless, and badly torn by the needlelike thorns, he scrambled down the tree to recover his undamaged rifle, which he had dropped in the undergrowth. Then he heard a scream of agony from Hamisi.

"I do not know what actually happened," records Foran, "but I think he must have about turned to run for a safe refuge —most abnormal behavior on his part. He was now some distance away. Scrambling erect, I saw Hamisi thrown to the ground and the elephant's massive bulk bestriding him; and then the bull began stabbing viciously with its long tusks, and trampling him savagely with the forefeet."

Foran ran toward the elephant. "At almost point blank range I put a shot under the root of the beast's tail, and followed this up with two more bullets into the same spot. This must have shaken the bull severely. I hoped one of the bullets might have raked the body to get the heart. It abandoned Hamisi, sped down the path, screaming shrilly, and tail well tucked in.

"Immediately the bull had gone, I knelt down beside the gunbearer to do all in my power for him. It needed only a glance at his terribly mangled and battered body to assure me that he was beyond human help. The elephant had gored and crushed him almost out of any resemblance to a human body. The wicked tusks had disembowelled Hamisi, both legs and arms were fractured in several places; and the spot where he lay, scarcely breathing, had been converted into a blood soaked shambles. Even the spare cartridges, carried in slots in his shirt front, were bent and twisted almost double. The angry beast had left behind not a human body, only a pulped mass of flesh and broken bones. I did everything possible to ease his passing. He died as I moistened his twitching lips with cold tea from my water bottle. His eyes opened and stared into mine. A faint sigh, a light and convulsive shudder, and Hamisi had ended his last safari and six faithful years of service with me."

Hamisi bin Baraka was buried within a short distance of the spot where he died. The hunter never saw the bull again. It was found days later, dead, and 175 pounds of ivory was taken from its head. W. Robert Foran never again hunted elephant.

This incident illustrates graphically the blind fury of a rogue African elephant, its fantastic ability to absorb punishment and its single-mindedness in attack. Of all big game animals a charging African elephant is the most terrifying and difficult to stop except with a really heavy rifle. Although many hunters tend to play down the dangers of elephant hunting, more of them are killed by elephants than by any other animal.

In many ways it was the intrepid ivory hunters who opened up Africa. In some areas their example and their influence upon the indigenous people was far more beneficial than the influence of the missionaries, who were often completely out of touch with the brutal realities of Africa. It was George Rushby, the hunter-author, who said that he was often more impressed by the little mounds in the veld covering the remains of some unfortunate ivory hunter than he was by the tablets in Westminster Abbey. Some great and gallant men were killed by elephants.

One of the saddest of all these deaths was that of Billy Pickering, a well-known and well-liked ivory poacher who bagged an incredible tusker carrying 190 pounds of ivory on each side. The next elephant he stalked tore his head from his shoulders. Mannix quotes F. N. Clarke, who had found Pickering's head some distance from his trampled body. Clarke says that according to Pickering's gunbearers the hunter had taken a bead on the elephant but for some reason did not pull the trigger. He just stood there pointing his rifle as the bull charged down on him. He seemed paralyzed. Oddly enough, the same thing happened to Clarke: he was once aiming at an elephant from a canoe when he lost his nerve and could not squeeze the trigger. Fortunately the elephant moved off. Clarke never hunted again.

Loxodonta africana, the biggest land animal on earth, is never an easy quarry although one would be forgiven for imagining at times that it it. John Taylor, for instance, shot four with four bullets using a medium-to-heavy caliber. He also shot 38 in a day and on another occasion, using a Mauser-action repeater, killed 12 with 14 shots at Mangwendi and then 7 more to completely annihilate a herd. But he never treated his quarry with anything but respect. Rushby, when he was in charge of elephant control in the Eastern Province of Tanzania, killed, with the help of African rangers, an average of 1000 elephants a year. He averaged 3.36 rounds of ammunition per elephant and that included shells used for training recruits. If this makes it look relatively easy, bear in mind that three or four of his African marksmen died each year, most of them killed by elephants.

Once an elephant does turn rogue it can chalk up an enormous death toll in an area where there might not be a weapon capable of killing an elephant or a man brave enough to hunt it. In recent years a marauder in northern Zululand killed twelve people one by one in a temporary fit of madness. It was never shot. Rogues like this have laid villages to waste, terrorized users of busy roads, and charged railway trains. The Dabi killer elephant, which frequented an area south of the Zambesi, is a typical case. Apart from his crop raids, he would ambush paths, chase victims, and then destroy them— usually by beating them against trees and then, placing a fore-foot on them, ripping off their limbs and scattering them. The Dabi killer was shot by John Taylor while it was raiding a millet field.

The "modern elephants" (those who have learned to change their habits and live in hiding where man cannot hunt them) are able to live in an area yet may never be seen by the local populace. This makes the sudden emergence of a rogue an even more terrifying event. At Ingwavuma, Zululand, in 1958, a twenty-year-old Zulu, Shanakeni Tembe, gave evidence at an inquest into the death of his mother, who had been killed by an elephant as she walked with Tembe. The African told the inquiry that he saw the beast rumble out of the forest beside the track and lift his mother off the ground before hurling her against the trees. He had no idea what the animal was, for he had never seen one or heard of one.

The custom of ripping a victim's body apart and scattering the pieces is common in attacks by rogues. They will occasionally do the same to a hunter's rifle. Another common action on the part of a rogue is to toss its victim into trees or straight over its back. Many men have survived this sort of handling. Among them is Murray Smith, who was tossed into the air by a rogue and fell through a tree to find himself lying at the elephant's feet. The beast thrust at him with its tusks, which dug into the ground each side of the hunter. Murray Smith lived only because the elephant's firmly planted tusks prevented the tusker from crushing the life out of him with its forehead. Some elephants will catch men in their trunks and then impale them on their tusks and yet others have the strange habit of burying their victims, or covering them with leaves. Some people have lived after feigning death and being buried by elephants.

Two hunters, the great Jim Sutherland and Samaki Salmon, in entirely separate incidents, went through the harrowing experience of being picked up by a charging elephant, tossed through the treetops, and landing on the ground near enough to their fallen rifles to pick them up and kill the elephant. Other hunters, pursued by determined elephants, have decoyed them by casting off items of clothing as they ran.

The African elephant can weigh up to twelve tons (Fenykovi shot one of this size in Angola in 1955; its skin weighed two tons) and it can measure 13 feet at the shoulder. The average bull elephant is 10 feet 6 inches. The African elephant has been known to carry almost a quarter of a ton of ivory, to eat a ton of vegetation a day, to carry three-quarters of a ton in its stomach and gut, to climb up to sixteen thousand feet—at times scaling slopes that would force a man onto his hands and knees—to live to be seventy, and to charge at fifteen miles an hour. Not only has this remarkable animal held its own in areas where other game has dwindled to almost nothing, but its number has increased throughout Africa to such an extent that now it must be continually thinned out.

But the elephant, although comparatively intelligent among the large mammals, is in fact not as bright as some try to make out. To control its massive bulk—remember it weighs far more than a double-decker bus—it has a brain the size of a cricket ball and at times it behaves as if it had no brain at all.

What turns a normally passive animal like the elephant into a rogue? Many are wounded by tribesmen attempting to protect their crops with such hair-raising weapons as ancient blunderbusses filled with nuts and bolts. (The biggest tusks ever collected were from an elephant killed by an African using an ancient muzzle-loader.) Perhaps 70 percent of rogues have been injured by guns, spears, or traps. Many others are suffering "legitimate" injuries received in interherd fights. T. Murray Smith shot one after it had charged him on sight and found 10 inches of ivory rammed into the roof of its mouth. Other causes are drunkenness brought about by eating fermenting marula berries (*Sclerocarya caffra*), through being mentally deranged (the incidence of insanity in elephants is as frequent as it is in humans, states Sanderson) or, on rare occasions, by having its tusks struck by lightning or otherwise damaged so that septicemia sets in. Rogues, according to Sanderson, do not just vent their spleen on humans; they also "lash out at little, inoffensive animals and rip up palm trees. They certainly seem to be 'batty.'"

There is little doubt that persistent hunting has made the elephant far more irascible than he was in the last century. Gone are the days when herds were found in the open and where a dozen or so could be picked off before the others fled.

The elephant of the twentieth century is very different and much more dangerous than the elephant of the nineteenth century.

In the African species of elephant (*L. africana*) both sexes bear tusks. There is evidence that the proportion of tuskless African elephants of both sexes is growing, probably because they have been ignored by ivory hunters and therefore live to breed. For some reason the tuskless elephants are worse-tempered than the tuskers, and far more caution is called for when they are encountered.

The best way, perhaps the only way, to save oneself from an elephant charge along a forest track is a brain shot with a heavy-caliber rifle. How accurate and how heavy the weapon should be is well illustrated in another of Foran's passages, one dealing with the hunt for Pepo, a rogue that was responsible for many deaths in Kenya.

"Suddenly with a shrill and infuriated scream, Pepo charged full tilt at Harry from about 30 yards behind him. Harry swung about at once, rifle to shoulder, and fired a bullet from his .460/.500 rifle into the forehead of the huge bull. It failed to kill. But the shock stunned Pepo and made him swerve slightly off course. Neither Jim nor I could shoot because Harry stood in our direct line of fire and with his back to us. The density of the undergrowth on either side of the track prevented a change of position to shoot with any degree of safety. Both saw and heard Harry make another frontal brain shot at the bull, but this second bullet was equally ineffective. Pepo just shook his massive head. Trumpeting loudly, the bull charged all out at Harry, who was reloading the rifle since his gun-bearer was not near enough for an exchange of weapons.

"A moment or two later Pepo was upon Harry. We saw him hurled violently to the ground, with the long tusks stabbing and ripping at his body; and then he was seized around the waist by the trunk, lifted on high and flung against the bole of a giant tree. A nerve-shattering scream of agony froze the blood in our veins; but a second later that dreadful sound was cut short. As Jim and I ran to his aid, I saw Pepo prodding savagely at Harry's prostrate body with the tusks and heavy forefeet stamping upon him."

From twenty paces Foran, using a .300 Rigby-Mauser—a gun that could virtually blow a man's head off—fired a frontal brain shot. It still did not knock the elephant down but at least caused it to flee.

If there is one animal in the world that, at close quarters, is an equal match for man and his rifle, it is a maddened elephant. ". . . when you hear a long, deep sighing sound . . . the elephant has had enough . . . when the sigh is followed by a short, shrill, high-pitched trumpeting blast, or a low, ominous rumbling, it means that the bull has pin-pointed the hunter and a charge is imminent. When the crashing charge comes, pushing over everything in the way, the moment of truth has arrived for both hunter and bull."

Both the African and Indian elephants have quite a range in sounds and among the most sinister of the warning signals is the tapping of their trunks upon the ground. When these hollow-sounding taps begin, the herd, if there is one, falls silent and even the constant stomach rumblings cease. An elephant is also able to emit a roar which puts fear into any man or beast.

When the male elephant charges it moves with the same sort of gait as a well-trained brewery horse, its massive feet pounding rhythmically (the female shuffles rather than runs). The head is usually held up and the trunk either dangles free or is tucked in under the chin. Carrington claims that the trunk is tucked in when the elephant charges in thick country but is held extended forward in open country. John Taylor describes the ears during a charge as "like a square-rigger's sails." The ears can span a good fifteen feet.

Elizabeth Balneaves describes a full-blooded charge by a cow elephant which was about to trample an African named Siamwanza. The African leaped behind a rock and the elephant crashed into the rock head on, shattering her tusk into a thousand fragments.

There is some truth in the delightful piece of advice offered by ivory hunters that if you stand stock still when charged by an elephant it might well miss you and thunder past. It might also lose sight of you and pull up. Some authorities claim that elephants are shortsighted, and such a ruse has saved quite a few lives. One cannot help wondering, though, how many it has cost. Men who have written their hunting experiences have assured us that even when a herd of elephants is bearing down upon them they have remained still, relying on the elephant's supposed instinct for going around any strange object. There is every reason to believe the hunters' story that men with faint hearts have died only because they moved at the last moment.

I have mentioned before the fantastic ability of the elephant to absorb bullets and the fact that the only certain way of stopping a charging elephant is with a brain shot using a heavy-caliber bullet. The heart shot, used in a tight corner, is usually a wasted one. George Rushby, hunting in a swamp with a .450 double-barrel, gave a passing elephant a perfect heart shot, which caused the elephant to double back along its tracks. Rushby says, "I was standing waist deep in swamp, some three feet to the side of his tracks, and my second barrel failed to turn or stop him . . . when he was nearly on me, I threw myself sideways, away from the tracks, but he scooped me up in his trunk in passing and held me up in the air. He carried me like this for a short distance then flicked me backwards. I fell on his hindquarters and rolled off into the swamp. The elephant continued on its way for another 30 to 40 yards and then fell dead." Elephants shot through the heart do not always die. Rushby records an elephant living three hours and walking a mile with a .303 clean through its heart. He also found a muzzle-loader missile in the heart of an elephant. A. L. Barnshaw, writing to Denis Lyell, described how he shot an elephant at point-blank range—straight into the heart. The elephant traveled on for four miles "and was ratty at the end." Barnshaw claimed that the hole in the heart was big enough for a fist.

Many African tribes successfully hunt elephant with spears and pangas. The art of killing the pachyderms with such crude weapons is an old one and Chaka, the Zulu king of the early nineteenth century, perfected it. He trained his impis to stampede elephants onto narrow mountain paths, where they could be ambushed by men armed with axes. The men would then rush out and hack at the thick tendons behind the elephants' knees. An elephant is unable to move on three legs, and the impis would then spear the animals to death with their hand-held stabbing spears.

About two hundred years ago the Arabs used to ride elephants down on horseback. Two men would be in the saddle and, as the rear man leaped off behind the elephant to hamstring it with an axe, the rider would distract it. Once the hamstringing was accomplished the rider would pick up the axe man and move him back to safer ground; then he would ride in and spear the animal to death. Occasionally horse and two-man team were torn apart.

Perhaps the most audacious method still in use is in the Congo, where some Africans smear themselves with elephant dung to mask their smell and creep up on resting elephants and then sink their spears into their hearts. The method must cost a considerable amount in lives. South African Bushmen have a safer method: they shoot them with arrows dipped in curare and then follow the spoor day after day if necessary until the elephant dies.

A significant account of the shooting of a bull elephant using old-fashioned muzzle-loading guns is contained in the diary of Jacob van Reenen (1790–91), who went overland along South Africa's Wild Coast in search of survivors of a wrecked East Indiaman. The account is an interesting example of the discouraging tenacity and slyness of a wounded elephant.

"Wednesday, December 1, 1890 . . . a big bull elephant appeared at the wagons, which we at once began to hunt. After he had received several bullets and had already fallen to the ground twice, he crawled into a thorn bush thickly covered with undergrowth. Tjaart van de Walt, Lodewijk Prins and Ignatius Mulder rode up to the bush thinking that the elephant had been finished off, when the elephant charged furiously from the bush and caught Lodewijk Prins, struck him from his horse's back with his trunk, and trampled him to death, and pierced him, Lodewijk Prins, right through the body with one tusk, threw him fully thirty feet in the air and away from himself. Tjaart van der Walt and Mulder, seeing no way of escaping with their horses, leapt from them and hid in the undergrowth. The elephant, seeing only van der Walt's horse, pursued it (they were all within an ace of being crushed to death by him). After having pursued the horse for some distance he turned again to the spot near to where the dead body was, looking for it. Then we all began shooting at the elephant in order to drive him away from the place or to shoot him dead, whereupon he finally, after a great deal of shooting, hid in the thick thorns. When we thought he was far enough away, we began to dig in order to make a grave for the unfortunate man, and while we were busy with this the elephant charged us again and chased away all who were remaining by the grave in the open plain. Tjaart van der Walt, once again riding past him at a distance of 100 yards, shot at him again. The fury of the animal was indescribable. We shot at him again with our entire force until at last he left the spot, and then let him be followed by the Hottentots who shot him to the ground once more, and he remained lying there."

They then continued with the burial.

There follows this note:

"N.B. All those with me, who were old travellers, had to admit that this was the fastest and most vicious elephant that they had ever seen, and the Hottentots told us that such an elephant has the habit of not leaving the body, but of swallowing it piecemeal; and that they had seen a Hottentot perish in this manner, and had been unable to discover any trace of him."

A possible explanation for the occasional disappearance of an elephant's victim is that he will sometimes bury the victim under dust, leaves, and branches. But the theory that the victim was eaten is interesting in the light of other stories indicating that elephants might, on rare occasions, be man-eaters.

In 1944 in the Zurich Zoo, Chang, an Indian elephant, was looked after by a typist named Bertha Walt, who was given special facilities at the zoo even to the extent of being allowed to sleep in a room next to the elephant stall.

One morning the keeper found her missing and then noticed blood on the stable floor. The keeper then found a human hand and a toe. Some days later, fragments of Bertha Walt's clothing appeared in the elephant's droppings.

Some years before this Zurich incident, at Pagel's Circus in the Transvaal, a drunken man emulated the ringmaster for the entertainment of some children standing nearby. He put his head in an Indian elephant's mouth. The elephant closed its jaws and lifted the man off the floor—then dropped him. His head had been chewed to a pulp.

Now an elephant that can process in its stomach each day a ton of coarse vegetation, including thick tree branches, should be able to dismember and successfully chew and swallow a human. The most surprising factor is not the size of the meal but the fact that it was meat, for the elephant is a vegetarian.

According to hunters in Africa, an elephant will occasionally carry a victim around in its mouth, and there are stories of elephants dismembering victims and carrying a limb around in their mouths. It seems feasible that they might occasionally chew and swallow these limbs.

The Zurich story is not unique. Carrington mentions the "Mandla" elephant, which terrorized a district of that name near Jubbulpore in the Central Province of India, and which, in the early 1870's, was said to have eaten some of its "numberless victims." This, says Carrington categorically, "is of course nonsense, as elephants are exclusively vegetarian in diet, but the superstition probably arose from the animal's habit of playing with the limbs of dismembered natives, and holding them in its mouth. The elephant was eventually shot." This "Mandla" elephant is no doubt the same as the "Mandala" elephant mentioned by Sanderson, who also suggests that the stories concerning its man-eating arose from its habit of carrying its victims in its mouth before tearing them apart.

Captive elephants in zoos and circuses—nearly always Indian elephants—have been responsible for a lot of deaths and sometimes appear to go through temporary periods of insanity. Most zoos in the world have banned elephant rides over the past ten years because of the number of people killed. It is odd though that if a captive elephant does kill somebody and its summary execution is ordered by the authorities, the public occasionally petitions for its reprieve. Usually pleas are for distorted humanitarian reasons but sometimes, especially in the East, they can be for economic reasons. If a trained work elephant kills its mahout or grasscutter it is usually rested for a while; almost never would such a rogue be shot. It is very difficult to come by a well-trained elephant in India (total elephant population, c. 7000), but handlers are ten a penny (total human population, c. 450 million). A certain Indian elephant killed "some eighteen men," but because he was a good worker he was saved from execution. This killer began his career by trampling his grasscutter. He gored the replacement to death too. "Shiv Dutt," as he was known, was eventually killed by a wild tusker. According to P. D. Stracey, the Raja of Bijni owned an elephant that even after killing a dozen men, one by one, continued to be used as a work elephant. Stracey has a soft spot for elephants: he relates how a five-year-old Florida boy who had been stealing pennies from a circus elephant named Dolly was trampled to death. "Poor Dolly," says Stracey, "was executed with cyanide."

The differences between Indian elephants and African elephants are marked. The Indian elephant is much smaller (average 4 to 5 tons), has smaller ears, and has a convex forehead and back instead of concave like the African elephant. The Indian female is tuskless (except in freak cases), and the tuskers of Asia carry far less ivory than the African elephants.

The nature of the Indian elephant is more docile and predictable, but whether the Asian species is more intelligent is another matter. It is unlikely, but the Indian elephant has certainly served man in the East far more than the African elephant has served the African. This difference, however, may be more the fault of the African people than of the elephants themselves.

Most men who have hunted both species do not consider the Indian elephant particularly formidable, although the solitary bull, according to E. P. Gee, can be dangerous. Gee also claims that there is no evidence that the bull Indian elephant is dangerous when in musth ("in heat"—the males have heats, not the females). Kenneth Anderson disagrees, saying they are "extraordinarily dangerous" at the time.

Nevertheless the species *Elephas maximus* can certainly produce formidable killers from time to time. Akbar, a remarkable and large riding elephant at Kaziranga, was often used by the Forest Department to carry tourists on its back. Few could have realized it was a man-killer with three kills to its name. This unstable giant twice put the fear of the devil into some tourists it was carrying by fighting a tiger to death; on a separate occasion, it fought and put to flight a wild elephant.

Rogue Indian elephants are apparently every bit as dangerous as rogue African elephants. Carrington estimates that fifty people die in India every year through their attacks. A rather bizarre example of a rogue elephant in India was described by George P. Sanderson in 1878 in *Thirteen Years Among the Wild Beasts of India*. Near the village of Kakankote fifty miles from Mysore, a bull elephant took to alarming people by charging out onto the road. Then it began killing people by putting a foot on them and ripping off their limbs with its trunk, scattering them widely in the manner of African elephants. Guards posted at each end of the eight-mile stretch warned people of the rogue, and travelers would play drums and horns in the hope of driving the mad elephant away. Sanderson hunted it and wounded it, causing it to lift its siege for five months. But it returned and in December 1872 he went after it for a final time. He managed to get in a heart shot but the tusker did not go down. Sanderson followed the blood trail into thick bush, a dangerous thing to do in the circumstances, and within a short distance became sticky with blood. Suddenly the beast was facing him, blood frothing from its mouth. He felled it with a shot between the eyes.

The first elephants used in war probably were trained in India well over three thousand years ago and were used in much the same fashion as tanks are today. The art of ancient India has left us many examples of the use of war elephants, including some which carried nests of javelin throwers; others carried towers from which soldiers could scale enemy fortifications. About this time, say the latter half of the second millennium B.C., the Chinese began making increasing use of war elephants. As the Eastern armies became more sophisticated, so the elephants were more highly trained and formidable in attack. Some were trained to wield swords in their trunks and had daggers lashed to their tusks. Some wore metal armor. Their psychological effect must have been great, especially when they bore down upon foot soldiers who might never have seen such creatures before.

The emperors of the East kept vast numbers of war elephants right up until the Middle Ages. In fact, the King of Rangoon had five thousand war elephants besides many untrained ones until about three hundred years ago. The advent of the firearm more than anything else made these living tanks obsolete, but the Indian army still employs ceremonial elephants and in quite recent history fired cannon from their backs.

In 255 B.C. the Roman army was the finest in the world. The regiment of expeditionaries commanded by Atilius Regulus was an excellent example of Roman fighting efficiency and discipline and was 15,000 strong when it prepared for battle on the Tunisian battlefield of Tunctum. The regiment was flanked by 500 horsemen.

A mile away and facing the Romans was the motley, mercenary army of the Carthaginians under the Macedonian Xanthippus. Carthage had already been badly mauled by the

Romans and its fleet recently defeated. The Carthaginian force was, in numbers but not in discipline, about the same as Regulus' force with the exception of horsemen, who numbered 4000. Regulus, in spite of the Carthaginian horsemen, was probably confident of victory even though the battle was likely to be hard fought.

The two great armies began a disciplined, even movement toward each other when, quite unexpectedly, the Carthaginians started a maneuver that resulted in two wide passages forming through its ranks. Down these passages, in single file, came the elephants. The effect upon the Romans can hardly be imagined, for none had seen elephants before. The massive, gray beasts, wearing iron armor about their heads, shuffled in their typically waltz-like gait down the lines and out across the dusty plain toward the Romans. They were African elephants, ten feet high at the shoulder and with ears spanning twelve feet. On their backs rode the mahouts, each armed with a spike and a mallet. The weapons were not for use on their enemies: the spike was to be driven between the base of the elephant's skull and the first vertebra should the creature run amuck in its own lines. The blow would collapse the elephant in a heap.

As the great pachyderms advanced, their trunks groping forward, the Romans fell back. Then the elephants were upon them, flailing them with their trunks, hurling them into the air and trampling them underfoot. The Romans broke and ran but few men can outrun the steady fifteen-mile-an-hour charge of an elephant. As the elephants battered down the Romans, the Carthaginian foot soldiers came in behind and cut down the survivors. Only five hundred Romans lived.

The Romans, after their initial defeats by elephant-equipped armies, were quick to find a countermeasure and within four years of the Tunetum débacle Metellus decisively defeated the Carthaginians at Panormus and even succeeded in turning the elephants back into their own lines. The Romans' resourceful, if unsubtle, answer to war elephants was to train squadrons of *velites* whose sole function was either to fire arrows into the elephants at close quarters or to stand their ground and spear the beasts as they bore down upon them. Others, armed with axes, hamstrung them. Blond states that catapults too were used, and the most effective projectiles were those covered in flaming tar or resin which stuck to the elephants' hide. The most diabolical ruse was for some *velites,* dressed in spiked armor, to allow themselves to be picked up so that they could lop off the elephants' trunks and cut into their heads or forelegs with their broadswords. The Romans no doubt found it easier and more economic to produce *velites* rather than trained elephants-of-war. (Hannibal, who lived about this period, was slower to learn. He created a legend when he took his elephant army across the Alps. Only one elephant survived.)

At one stage the Romans shipped 140 captive enemy elephants to Rome, where some were made the object of ridicule and taken in chains from village to village. The idea was apparently to dissipate the awe that the elephants created. The creatures were later led into the arena, where they were killed by lions, tigers, and men.

According to Pliny, elephants made their first appearance in the arena in fights against each other but the Rome mob soon tired of this and for a time wild cattle were pitted against them. But even the sight of bulls being lifted from the ground and torn apart failed to amuse the Colosseum's habitués. Inevitably men were pitted against the elephants. For a time death by elephant for offenders was, in more senses than one, the popular form of execution in Rome.

People were staked out in the arena while mahouts steered specially trained elephants to walk over them. Occasionally the men under sentence were allowed a spear or sword but the idea could only have been to prolong the agony and entertain the spectators.

Once the elephants attempted to get at the spectators by smashing down the barriers. Cicero, writing of this day, records that the crowd was moved by pity for the elephants. According to Dio, the crowd became so angry at the injuries suffered by the elephants that the tournaments were suspended. But, writes Ivan T. Sanderson, elephants came back to the arena and

Julius Caesar, curious to see how they might fare in war, watched a number of his countrymen kill and be killed by elephants.

A lone man, one might suppose, would certainly die in a duel with an elephant especially since most of these appeared to be African elephants, which are much larger and a great deal more short-tempered than the Indian species. On some occasions, however, gladiators were able to hack at the tendons in the elephant's back legs, immobilizing it, and then hack at the trunk, which is extremely sensitive. The gladiator would then attempt to drive his sword into the heart.

Pliny tells how Hannibal once forced two Roman captives to fight each other to the death. The survivor was promised his freedom if he survived singlehanded combat with an elephant. The Roman fought and killed the elephant and was then released. He never reached home. Hannibal, realizing that the Roman's achievement might lead the Romans to lose their fear of his war elephants, had the soldier murdered.

The Romans were never overly impressed with the elephant's qualities as a fighting unit, in spite of any fears they may have had of it. If they showed any interest in its potential, then it was the African species that they preferred. These were obtainable from the deep forests of North Africa—forests now buried beneath desert. On the one or two occasions when the Romans did use elephants in the field their successes were notable. Blond records that it was Scipio's Rome-trained elephants that were used to defeat seventy thousand Carthaginians in the battle of Nepheris, which paved the way to the sacking of Carthage. It is ironic that Roman elephants should have finally defeated the very people who demonstrated to the Romans in the first place just what elephants could do. The Romans predictably arranged their war elephants in well-disciplined phalanxes of sixty-four and "elephantarchies" of sixteen. A "zoarchy" was a single unit of elephant, saddler, cook, and crew. The crew would sometimes drug the elephant before battle or work its spirit up by giving it wine to drink.

Once the Romans had consolidated their victories, they insisted on "disarmament" as far as elephants were concerned and, having made sure there were no more war elephants in nearby countries, they retired their own. The golden age of war elephants was over, but in the Far East they continued to be used.

Marco Polo provides us with a useful account of the battle between the Tartars and a Burmese king in which the Tartars showed themselves to be far less dismayed by their first sight of elephant than had been the Romans some fifteen hundred years before. Tartars, mounted on their fast little ponies, were suddenly confronted by massed elephants carrying wooden towers. They fell back to a forest and dismounted. As the elephants charged, the Tartars released a cloud of arrows. Soon the elephants resembled giant porcupines and in their frenzy charged back into their own lines.

Muhammed Casim Ferishta in his *History of Hindostan* (c. 1600) gives an account of one of the last great battles in which elephants played a vital part: the battle of Chitor, where "300 elephants of war were ordered to advance to tread [the Rajaputs] to death. The scene became now too shocking to be described. Brave men, rendered more valiant by despair, crowded around the elephants and seized them even by the tusks, and inflicted upon them unavailing wounds. The terrible animals trod the Rajaputs like grasshoppers under their feet, or winding them in their powerful trunks, tossed them aloft into the air, or dashed them against the walls and pavements. Of the garrison, which comprised 800 Rajaputs and 40,000 inhabitants, 30,000 were slain."

Apart from their role in war, elephants have also been used from time to time as executioners. There is a seventeenth-century account by Edward Terry, whose evidence must be accepted with reservations, for he declares that "the male's testicles lye about his forehead." Terry spent three years in Great Mogul's court in India from 1619 when, he claims, the Mogul's elephant herd numbered 14,000. Although this herd would need at least 4000 tons of fodder a day it is quite possible that it did number 14,000.

"They are the most docile creatures . . . some elephants the king keeps for the execution of malefactors; who being brought to suffer death by that mighty beast, if his keeper bid him despatch the offender speedily, will presently with his foot push him into pieces; if otherwise he would have him tortured, this vaste creature will breake his joynts by degrees one after the other, as men are broken upon the wheele."

Carrington quotes Paul Edward Pieris Deraniyagala as saying that elephants (obviously Indian) were used in the East as executioners. They were trained to pick victims up in their trunks, place them upon the ground, put a foot on them and pull them apart. They were also trained to pull two trees together while a man was tied between them. The elephant would then let go and the victim would be split asunder.

The Rhinoceros

The black rhinoceros, *Diceros bicornis,* is the most bad-tempered animal in the African bush. When approached it is the animal most likely to charge—whether one is on foot, on a bicycle, on a horse, in a car or a truck. In East Africa they occasionally charge trains.

Is the African black rhinoceros really the most dangerous of all the big game animals? There are experts who say yes and there are men of equal experience who say it is the least dangerous, putting the lion, leopard, elephant and buffalo before it.

Although it is true that almost all the world's animals have to be sorely provoked before they will attack man, in the case of the black rhinoceros provocation can be the slightest of sounds, including the click of a camera. A gentle movement or a strange scent—or even a bird rising out of the grass— is sometimes enough to cause it to lower its ugly head and charge.

There is a theory held dear by a surprising number of big-game hunters that one can side-step a charging black rhinoceros and it will then go trundling past and eventually stop and begin browsing again. One of our best hunter-authors states: "The black rhino can be side-stepped with comparative ease." This is a fallacy: the one-and-a-half-ton rhinoceros can turn on a dime. Ian Player, Zululand's chief nature conservator and one of the world's top experts on rhinoceroses, told me: "I have known several Africans to be killed by black rhino over the years and I doubt whether there is a game ranger in these parts who has not been charged by one. The only sensible thing to do when confronted by a charging rhino is find a tree and climb it—even four feet from the ground is usually safe enough. You can forget trying to side-step a rhino. The best trick is to chuck your hat or bush jacket—or anything—in its way and hope it will take it out on that."

C. A. W. Guggisberg, the East African wildlife writer, correctly states that a number of people who claim to have been charged by black rhinos have merely been the subject of "an exploratory advance." If the black rhino senses something suspicious, he will throw up his head and trot toward the source of his annoyance. He might then trot around in a half circle and test the wind or he might retreat, turning now and then to face the direction where he suspects somebody is standing. It is best to assume that he is going to attack, and therefore you would do well to look for a tree to climb. Occasionally a shout or violent action will cause the rhino to rush off in a state of high alarm. Then again, it might make him charge. No other animal is so magnificently unpredictable.

It is my experience that a black rhinoceros will almost always advance if he suddenly senses you are near. Although he seems to have extraordinarily poor sight he has very acute hearing and a sense of smell. If he detects a slight movement or catches your scent he will more likely than not charge with lowered head and when a one-and-a-half-ton animal comes toward you at twenty to twenty-five miles an hour he is unstoppable. At the end of his charge he usually hooks left and right with his horns, and many a hunter bears the scars from such an experience. Unlike the buffalo, the rhino will often be satisfied with tossing a man high into the air (one can reach a height of twelve feet, according to Captain C. H. Stigand, who once took such a journey), but just occasionally he will whip around and then bore his victim into the ground or even gather him up on the end of his horn and toss him again and again. Eastwood, chief accountant for Uganda Railways in its pioneer days, had a remarkable escape when he approached a rhinoceros that he thought he had shot dead: the "dead" rhinoceros rose to its feet and then fell on him, cracking four of the man's ribs and breaking his right arm. Then it impaled him through his thigh and threw him high into the air. Twice more it tossed him. Eastwood was alone at the time and left groaning in the long grass. Had it not been for his African assistants, who saw vultures circling the spot, he would have died. His arm had to be amputated.

Lieutenant Colonel J. H. Patterson tells an exciting tale of a black rhinoceros that he saw in East Africa. He stalked it to within fifty yards of where it was resting and lay down in the grass to watch it. After a few minutes the animal began to suspect his presence, for it rose and walked around in a half circle trying to pick up his scent. Patterson recalled: "The moment he got wind of me he whipped around in his tracks like a cat and came for me in a bee-line. Hoping to turn him, I fired instantly; but unfortunately my soft-nose bullets merely annoyed him further, and had not the slightest effect on his thick hide. On seeing this I flung myself down quite flat on the grass and threw my helmet some ten feet away in the hope that he would perceive it and vent his rage on it instead of me. On he thundered, while I scarcely dared to breathe. I could hear him snorting and rooting up the grass quite close to me, but luckily for me he did not catch sight of me, and charged by a few yards to my left." Patterson then rose and very stupidly sent a couple more bullets after the animal. They disintegrated against its thick hide and caused the animal to stop short in its tracks. He began to gore the ground viciously, for he was too blind to see Patterson standing there; then he began to move again in a semicircle. "This proceeding terrified me more than ever," said Patterson. "I could scarcely hope to escape a second time." The rhino once again picked up the hunter's scent and "down he charged like a battering ram. I fairly pressed myself into the ground, as flat as ever I could, and luckily the grass was a few inches high. I felt the thud of his great feet pounding along, yet dared not move or look up lest he should see me. My heart was thumping like a steam hammer, and every moment I fully expected to find myself tossed into the air. Nearer and nearer came the heavy thudding, and I had quite given myself up for lost, when from my lying position I caught a sight, out of the corner of my eye, of the infuriated beast rushing by. He had missed me again!"

Patterson recorded another attack—surely one of the most devastating charges by a black rhinoceros. The animal charged out as twenty-one slaves who were attached by their necks to a long chain filed down an East African bush track on their way to the coast. The rhinoceros impaled the center man and the sudden jolt broke the necks of the others.

Whenever I see a black rhinoceros I recall Murray Smith's words: "It is a miracle that this prehistoric idiot still exists." It is indeed and it is conceivable that had the rhinoceros had better eyesight it would have been an intolerable animal to have where there are people. But then, I suppose, it could be argued that if it were not so poor-sighted it would not be so bad-tempered. Its bad temper and aggressive habits when man approaches are probably its salvation in another way: it has saved the animal to some extent from the ruthless African poachers who are paid one pound sterling for every pound weight of rhino horn by Asiatics who believe (quite wrongly) that it has aphrodisiacal powers.

Most big-game hunters agree that the black rhinoceros is easily killed with a medium or heavy rifle. Often it can be effectively turned off its course during a charge by firing over its head or into the ground ahead of it. It is certainly easily felled when it presents a head-on aspect. J. A. Hunter would allow them to charge to within fifteen yards of his clients' cameras before felling them with a single shot.

Perhaps if the white rhinoceros, *Ceratotherium simus,* had the same bad temper as its cousin it would not now be on the verge of extinction. It is a great deal bigger than the black, its three tons making it the second biggest land mammal on earth. The trouble with this enormous animal is that it is as docile as a Jersey cow; I have often stalked them to within a few feet. According to Ian Player, they have killed only four people in living memory in Zululand, where at least half of the last two thousand white rhinoceroses on earth live. All four delinquent rhinoceroses had been badly wounded and were driven to desperation before they charged. One had fifty spear wounds.

If historians are correct and the Indian rhinoceros was used in war as a sort of front-line tank by the kings of ancient India, then it does appear that *Rhinoceros unicornis* is more intelligent than he looks. The Indians are supposed to have lashed tridents to the horns of the rhinoceroses, which suggests they must have been trained. It is hard to believe, for the Indian rhinoceros with its single horn and heavily folded skin is every bit as stupid as the black rhino of Africa. Its method of attack is interesting in that it tends to use its lower teeth, instead of its foot-long horn, as an offensive weapon. Although it is said to be as aggressive as the black, I have found no cases where it has killed a man, but I don't doubt for a moment it has. Frequently the Indian rhino has put hunting elephants to flight. Gee says that annually the animal kills "a few people." Probably not many more than six hundred survive today in their strongholds in Assam, Bengal, and Nepal.

There are two other species in the East, the Asiatic or Sumatran rhinoceros, *Didermocerus sumatrensis,* which has two horns, and the Javanese rhinoceros, *Rhinoceros sondaicus,* which is one-horned. Both are practically extinct and both are said to be short-tempered and aggressive, and there are a few accounts of men being tossed by the latter but no actual fatalities. The Sumatran rhinoceros is said to be very aggressive and natives are afraid of it.

The Buffalo

"But this was Africa, and all things were
ominous."

Alan Wykes

Being hit by a charging buffalo is tantamount to being hit by
a medium-sized car. The beast usually weighs more than half
a ton—often a lot more. Some bulls can weigh well in excess
of a ton. Perhaps it would not be so bad if the buffalo were,
like the rhinoceros, satisfied with just knocking a man down.
But a buffalo will either hook a fallen man high into the air
(some hunters have landed in the tops of thorn trees ten to
fifteen feet above the ground, out of harm's way), or it will
churn him into the ground by first going down on its knees
and then goring the victim. It might then rise and trample the
dead man until he is almost a pulp.

The African buffalo must have given the pioneer hunters,
with their inefficient weapons, a pretty rough time judging by
the lurid accounts of their adventures. In fact the buffalo took a
fairly heavy toll of these early hunters and their entourage,
who, more often than not, had to have several shots before
the beast succumbed. One wonders from looking at some
accounts that the sheer weight of lead did not drag these
animals to their knees.

Wherever the pioneers went they left a trail of wounded
buffaloes, which so long as their wounds permitted them to
live were a terrible menace to the local populace. Following
in the wake of the explorers came the traders, who supplied the
natives with guns—mostly antique muzzle-loaders and deadly
blunderbusses (deadly, that is, to the user)—which meant an

even bigger population of wounded buffalo. The human toll rose accordingly and in some areas the buffalo was killing more people than even the crocodile. Many hundreds of people must have been killed each year by buffaloes throughout Africa around the turn of the century. Some of these rogue buffaloes are known to have killed a dozen and more people and tossed several others before they were finally killed.

F. C. Selous had some unhappy experiences with buffalo and, as much as anybody, he helped establish the animal in the public mind as a ferocious beast which was bent upon destroying man. The image persisted until quite recently. T. Murray Smith asserted, "The buffalo is not only dangerous, he can be the most wantonly vicious animal in the wilds." Cherry Kearton, who made a fortune filming and photographing big game, enhanced its reputation for charging on sight and tossing people among the treetops. "The buffalo," he wrote, "is far from being one of the safest animals for the unarmed man to photograph." Foran placed it as the third "most dangerous wild animal" in the world—next to the elephant and the tiger.

Syncerus caffer, the Cape buffalo, unquestionably is among the most dangerous animals when it comes to killing hunters: it might even rate second to the elephant. Unless one is armed with a medium to heavy, reliable rifle, it would be very dangerous indeed to try to shoot a buffalo. When wounded it can be a most terrifying animal even to a hardened hunter. But an unmolested buffalo is, if anything, a bit of a bore.

In the game reserves of Zululand guides take tourists on foot right up to herds of buffalo. Picnic lunches in their shoulder bags, the tourists click away with their cameras from fifty yards and sometimes even nearer. The guides carry rifles but they have never had to use them. I have often watched tourists creep up to buffalo herds, their hearts in their mouths, their f-stops forgotten. But when they are standing in front of the great black beasts with their leathery, padded noses and heavy bosses the tourists' reaction is usually disappointment. Buffaloes, grazing or browsing in a herd, look very much like a herd of rather large cattle. One or two in the herd might advance a few steps out of curiosity, nostrils aquiver and heads held high, but if one of the tourists makes a sudden movement or if the buffaloes catch wind of them the herd will wheel around and trot off to what they consider a safer distance. The buffalo, alone or in a herd, is normally shy and as nervous as an antelope.

Even when hunted, this animal need not be a dangerous adversary—not if the hunter knows what he is doing. John Taylor said, "Men get queer ideas about buffalo. Most men without much experience seem to think that buffalo will attack without any provocation. There is a belief that buffalo will invariably whip around on feeling the lead and make a savage and determined charge. Well, all I can say is that I have never experienced either of these things, and I have shot close to 1200 buffalo . . . and have encountered a hundred times that many." Eric Noble, a big-game farmer on the Rhodesian Sabi, ranches several hundred buffaloes as beef stock (as do many other farmers throughout southern Africa) and maintains that they are more tractable than Afrikander cattle. Foran, who once observed that the buffalo's "reputation for inherent savagery in its general disposition is unwarranted," records an incident which, to say the least, underlines the docility of unmolested buffalo. In the Tana River region of Kenya, Foran and his assistant Hamisi spotted a magnificent bull behind a large herd of buffalo. Herd bulls, says Foran, place themselves on the upwind side of the herd as it is normally from the downwind side that their one and only natural enemy, the lion, attacks. Foran and Hamisi tried for some hours to get around the herd without being detected, but late in the afternoon they found themselves back where they started. Foran so badly wanted the trophy that he decided to risk what he rightly calls "foolhardy action." He and his assistant began trotting toward the herd. The buffaloes nearest them raised their heads but showed no inclination to run. They obviously had not caught the man's scent yet. Foran and Hamisi, like a pair of football tackles, passed right through the herd, handling off any buffaloes that tended to get in the way. Once at the rear of the herd, the hunter picked off the bull. Only then did the herd stampede. Foran warned, "I do not recommend others to imitate such foolishness."

A belief that dates back well into last century is that a buffalo, when shot and wounded, will flee for a few hundred yards or even miles. Then, when the country affords it some concealment, it will double back along its tracks and wait in ambush for the pursuing hunter. It allows the hunter to pass it; the buffalo will then charge from the rear. From this alleged habit the buffalo has earned itself an undeserved reputation for cunning. The buffalo does, in fact, use a similar strategem, but if it has ever purposely doubled back to wait in ambush it must be counted as exceptional behavior. When a buffalo is shot its first instinct is to run. The late Owen McCallum stated, "A wounded buffalo will move on as you come up with it." He once wounded a buffalo and five times got almost near enough for a shot, but each time the animal fled. Once the animal feels it is cornered it will turn and charge in the most determined manner. The buffalo's behavior when it reaches thick bush may have given rise to the theory that it doubles back. Like most other wounded animals it will then rest, feeling relatively secure. The hunter, following the blood spoor, is at a dangerous disadvantage and might even run into the animal before he sees it. Many hunters have died in such circumstances.

Nothing is quite as determined as a charging buffalo, and once the beast has made up its mind to charge, nothing short of death will stop its rush. The only exception to this seems to be when a herd is "charging," or, more likely, stampeding toward the hunter. This infrequent situation is more tricky than desperate. Foran claims he prefers "mass charges" to individual charges made by wounded animals. Few hunters are killed in these stampedes, and it is Foran's theory that the herd is not knowingly charging the source of danger but is rather seeking a tactical advantage by rushing upwind. Even when a herd is rushing away from the hunter, it might suddenly wheel around and come straight at him. Foran says that a couple of bullets placed at their feet will often induce them to turn.

There is an alternative measure, one used by American Indians when faced by a stampede of bison: kill or down two or three in the same place. This tends to split the herd and send them glancing off in two directions.

Shooting up a herd of buffalo is not as dangerous as it seems, and because of buffalo population explosions in many parts of Africa this wiping out of entire herds is going on almost daily. John Taylor gives us an excellent example of what it is like when he describes a night shoot in the Zambesi River valley, where buffaloes were destroying crops. He approached the herd to within twenty yards or so, and using a pair of doubles as well as a powerful hunting lamp, he opened fire on the herd.

"My first shot slammed through the shoulder of a very big bull, dropping him instantly. They were still nearly all broadside on to me. At the shot they all swung and stood looking directly towards me. So my next shot was a frontal brain-shot; and I knew that most if not all subsequent shots would be similar. The second beast dropped in his tracks and I exchanged rifles. It was now a case of picking a target and firing just as quickly as I could, swinging on to another, firing, and exchanging rifles with my gunbearer. Both weapons were fitted with ejectors which slung out the fired shells, so that my bearer was usually able to have my second weapon ready when I needed it. . . . After my third shot the entire herd commenced edging towards me, jostling one another and crowding together. . . . Eventually I had the herd within ten yards of me, and still I continued to shoot. But now I took three or four paces backwards and away from them as Saduko whispered that a number of the buffalo were working up on my right. If I allowed them to get around too far they would get our wind. So, having taken up a new position, I swung around to face those and opened fire on them; and then swung around again and dropped another two from those in the center." The massive leader of the herd then charged into the center of the pool of light and wheeled away with the herd, now in full flight, following on his heels. Taylor squeezed off two more shots as they ran. There were no wounded. But there were twenty-two dead buffaloes—all within a few yards of one another. This was at night, of course. I doubt whether such a feat would be possible in daylight.

Although Taylor, and people like Charles Goss, who shot two magnificent bulls stone dead with a single .600 bullet, make buffalo-shooting look easy a seemingly good shot can turn into a nightmare. John Burger, the South African hunter, once nailed a large bull buffalo with a perfect heart shot using a .404 soft-nose. The bull, fortunately running away from and not toward the hunter, covered 187 yards before dropping dead. On opening the animal the hunter found its heart was just a pulp. An acquaintance of mine was killed by a buffalo that he and another hunter had hit eleven times. The first shot was a bad heart shot at two hundred yards. The buffalo then turned around and looked along its nose toward the hunters. The buffalo, having good eyesight, soon spotted them and charged full bore. It was then hit with the second barrel of the .270 (a favorite rifle among old-timers but for buffalo a little risky). The buffalo showed no reaction and continued its charge as the bullet buried itself in its shoulder. The hunter swapped guns and rashly pumped off two brain shots, which he could hear ricocheting off the heavy boss that protects a great deal of the buffalo's head from the front. He then—at one hundred yards —slammed a shot into the animal's right shoulder, and it went down for the first time. In an instant it was up and charging again. He then raked it with a sixth shot. His companion, who was carrying a .333, put the seventh shot in the animal's right shoulder, breaking it. The buffalo went down but again rose quickly and charged. Another shot hit the same shoulder and the buffalo stumbled but still came on. The hunter, who had started it all, tried a spinal shot through the neck, missed badly at about twenty yards, and had a near miss at about fifteen yards, but this shot brought the animal down again and it struggled to regain its feet. Then it came on again. The second hunter put a shot through its chest, which felled it instantly. It was a perfect frontal heart shot and the two men, both a little shaken, for the buffalo was lying but ten paces away, shook hands. The first hunter walked over to his trophy and placed his foot on it for the camera. The buffalo lurched to its feet, knocked the man down, pummeled him into the ground, and then fell dead. The hunter died instantly.

That is the one merciful thing about being killed by a buffalo: death is usually instant. This was the case when one-armed George Gray died. His lack of an arm was no handicap until the day he fluffed a shot at a buffalo which charged him. His brother had been killed in 1911 by a lion—also following an unlucky miss.

Murray Smith described the damage inflicted on a fifty-year-old African, who had been gored and trampled "literally into the ground" by a buffalo. His intestines had spilled from a hole ripped in his stomach, every limb was broken, and one of his arms was all but severed from his body. The man's face was pulp and his right hand still clutched his spear, the head of which was missing. The late Ken Beaton, chief warden of Queen Elizabeth Park, saw just how determined a buffalo can be when one charged his Land-Rover, ripping off a door and gashing the body in a number of places.

There are on record several "miraculous escapes" involving the buffalo. Once, a wounded buffalo, also in Queen Elizabeth Park, charged a mother and her child. The buffalo charged the woman first, hooked at her, and missed. It kept going and then caught up with the child, which it also tried to hook. Again it missed. It continued its flight until it was out of sight. Tobi Rochat of Acornhoek in the Transvaal, in 1966, was charged by a wounded buffalo. When he looked for his gun, which was in the hands of his gunbearer, he saw that the fellow had fled. The buffalo rammed Rochat (who was in his sixties) into a rooibos, which is a rather springy type of shrub. Every time the beast tried to push the man into the ground the bush took up the shock and Rochat bounced back like a jumping jack. The buffalo gave up. In Uganda once an

African riding a bicycle was charged by a buffalo and managed to bang his hat over the animal's eyes. The buffalo stopped to shake it off and the African got away.

But normally it is practically impossible to avoid a charging buffalo for, unlike so many other animals, including the domestic bull, it keeps its eyes on you all the time and only at the last minute does it lower its head.

It is difficult to say how many victims the buffalo chalks up annually in Africa—probably in scores rather than in hundreds.

The Indian (or Water) buffalo (*Bubalus bubalis*) has never been regarded as particularly dangerous, although wild ones are potentially dangerous when wounded. The gaur (*Bibos gaurus*) "can be bad tempered and even dangerous," according to Gee, but it has no record of killing men.

On the North American continent there are two potentially dangerous bovines: the bison, which has no record to suggest that it is anything more than an accidental man-killer, and the musk ox, which again has the potential but is slow and incredibly easy to shoot and has no significant record to suggest it is a man-killer. Pedersen says that the musk ox is indecisive when approached, and this indecision is usually fatal for the animal. He writes, "It is really no match for a man."

Although the buffalo is a potential killer, and although it destroys crops throughout its range, causing death and misery through starvation or at least malnutrition, this same animal might one day make up for all that. Conservationists see in the buffalo a potential beef animal that can be raised like ordinary stock but with many decided economic advantages. In the bushveld it takes eleven years to improve the land until it is fit for grazing cattle, and even then one has to battle continually against bush encroachment and various tropical diseases which hit cattle in Africa. But with buffalo the bush need not be improved, bush encroachment is no problem, and the buffalo utilizes vegetation better than cattle. In fact, where a cow is only 8 percent efficient at turning the protein in grass into protein fit for human consumption (meat and milk), the buffalo is 40 percent efficient. It is also disease-free, except for rinderpest, and can be culled at the rate of 25 percent a year. Seventy-five percent of the carcass is edible. It is astonishing that its potential as a food animal has not been fully realized by farmers in Africa. Half its trouble, possibly, is its dreadful reputation.

The Hippopotamus

"There is no animal I dislike more than
the hippo."

Samuel Baker, explorer

It was April and the Zambesi was still deep and swollen from
the summer rains. The moon rode high and it picked out the
quietly running river sliding like quicksilver between the
shadows of the heavy vegetation on either bank. The high-
pitched ping of a fruit bat hung on the still air, and somewhere
in the reeds a crocodile moved and sent silver ripples pulsing
out of the blackness along the river's edge.

Then, faintly at first but growing steadily louder, came the
rough whine of an outboard motor. Bryan Dempster and his
two assistants, Joseph and Albaan, were returning from a successful crocodile hunt with three good skins aboard. It was
not quite four in the morning.

Dempster decided to cross a quiet pool where the river
ate into the riverine bush. There was no warning: one moment
the dinghy was heading across the smooth water leaving a
perfectly straight, silver wake and the next moment it was
flung clear out of the water and was rolling over in the air.
Guns, the lamp, the crocodile skins, and the three occupants
fell from it as the engine, tearing uselessly at the air, screamed
and then cut as it crashed back into the river. Dempster caught
sight of a massive bull hippopotamus submerging. And then
as the hunter lay floating on his back he saw the gigantic head
reappear and the great ivory tusks glint in the moonlight as the
hippo's jaw clamped over the boat and crushed it to matchwood.

Dempster knew that his only hope was to remain as still as possible, for any movement or sound would give his presence away to the beast that now floated, eyes and nostrils just above the surface, only a few yards away. Even if the hippo missed him the crocodiles might not. Silence flooded in again when suddenly Albaan began screaming. He couldn't swim, Dempster remembered. The African screamed and shouted and set up a frenzied splashing and Dempster knew that the African was doomed and so would he be if he reacted to the man's calls for help. He had to grit his teeth to keep himself from shouting. Once again the hunter saw the hippo's monstrous head emerge, and he saw the great jaws open and shut. Albaan's scream was cut short.

Death in the African bush comes in many different ways, but mostly suddenly and violently for man and animal alike. It is not like in Europe and America, where men die discreetly in their beds and where the family doctor clicks his bag shut and then quietly leaves the room. In the African wilderness death is rarely dignified, and almost everywhere one finds little mounds of earth and perhaps an unofficial cross. Usually nobody knows who lies buried there or even who buried him but they usually know how he died. If the grave is found along the banks of the Zambesi Valley it would normally be a hippo that killed the victim—perhaps a crocodile, although this reptile rarely leaves anything worth burying.

Hippopotamuses kill a surprising number of people in Africa each year; judging by the number killed in a relatively well developed country such as South Africa, the over-all figure for the entire continent must run to a couple of hundred a year.

The majority of hippo victims are killed after being flung from boats or dugout canoes. One of the worst areas is the Murchison Falls, where the Bugunga tribe hunt crocodiles from dugouts. These tribesmen, typical of most Africans, cannot swim.

In March 1959 a young South African Bantu Administration official, Andries Steyn, was ordered to shoot a rogue hippo that had been upsetting dugouts on the Okavango River. He shot the rogue, from a boat, but the bang excited a cow hippo, which emerged under the boat and threw Steyn out. She then attacked Steyn and dragged him under. He was never seen again.

Not all attacks are in rivers. Quite a few are on the riverbank or even well away from the river. Many a hunter has learned the hard way that hippos hate fire and will sometimes charge a campfire with disastrous results. Few animals are as unpredictable as this monster.

On the banks of the Pafuri near the Makulika trading store in the northeast Transvaal an African was walking along a path with his wife, who had a child on her back. The African saw a bull hippo and just as a precaution he shouted at it, knowing that it would then most likely plunge through the reeds into the river. But the hippo took it into his head to charge. It rumbled up to the man's wife and with two quick bites severed the woman's leg at the hip and took a large piece out of her side. She died minutes later. Her baby was unhurt.

In 1966 a hippo on the Limpopo, near Messina by the South Africa–Rhodesia border, took exception to a noisy party of Africans cavorting on the riverbank, charged out with mouth agape and bit one of them clean through the torso, killing him. It was the third fatality on that section of the river in a year. In 1961 an African child near Charters Creek, Zululand, was bitten in half by a hippo which charged from the water. Hippos, in spite of their rather peculiar dentition, are able to chop animals in half. One was seen snapping a ten-foot crocodile in half on the Pafuri. In 1964 another South African Bantu Affairs official, William Steynberg, was taken by surprise as he stood on the banks of the Pafuri. A hippo surged out of the river and bit him once but that one bite was enough to stave in his chest, exposing his heart and lungs. He died after a nightmare car journey through the bush to Messina.

Once I witnessed an exceedingly narrow escape by Piet Barnard, a hunter in the Olifants River, Transvaal, when a hippo charged him as he stood in the shallows. The hippo, a very young one weighing perhaps a ton, turned its head sideways for the first bite and its teeth bounced off the man's hips and left a deep, red weal across his stomach. The second bite crushed his elbow and the third pierced his armpit but miraculously did not remove his arm. The hunter flung himself backward to enable a colleague to empty both barrels of an "elephant gun" into the hippo, killing it even before it sank. In Barnard's case the incident was triggered by an unusual situation (he had been using experimental capture drugs on the hippo), but this type of accident is often the result of the rivers being shallow because of drought and the hippos feeling insecure.

Death from hippos occurs frequently when Africans are caught on hippo paths at night. The two-and-a-half-ton hippo is a grazer and leaves the river at sundown to forage along the bank and even deep inland. Invariably, year after year, it uses the same old track and when disturbed it will turn back down the track and race for the river, steadfastly refusing to be put off by any obstacle that might present itself. Many men have died because they did not jump out of the way fast enough.

Willock, who describes the hippo as among "the most aggressive and dangerous game in Africa," tells the story of one that drove its great ivory tusks through the side of a Land-Rover.

But as with most animals there are two sides to the story. The hippo can be aggressive, but when one considers the massive hippopotamus population of Africa (in some parts the population is outstripping the food supply), the incidence of "sour hippos" is relatively rare. Ironically, but perhaps typically, where the hippos have been exterminated the local African population suffers in an unexpected way. Hippos, as I have said, tend to use the same tracks when leaving the river and those tracks tend to go away from the river, always in the direction of the river's flow. Thus, when the floods come and the river swells, it gradually fingers out along these deeply compacted paths. When the hippos have gone and the Africans are able to sow their crops right down to the river's edge, the paths are ploughed up. When next the river rises it is no longer able to seep gradually into the plain: instead it rolls up the countryside and great quantities of topsoil are carried down to the sea. Within a short time, where for centuries hippos have grazed, the grassland is gone.

III

MAN-KILLER BEARS, WOLVES, AND HYENAS

The Hyena

The Bear

"When the Himalayan peasant meets the he-bear in his pride,
He shouts to scare the monster, who will often turn aside.
But the she-bear thus accosted rends the peasant tooth and nail
For the female of the species is more deadly than the male."

Rudyard Kipling

For hours El-Kalak, the Eskimo, had sat, his gun across his knees, watching the distant polar bear cavorting on an ice floe. The animal was within easy rifle shot but El-Kalak did not fire. "Mountie" Montagu could contain himself no longer. He whispered to the hunter, "Why don't you shoot?"

"One awaits better luck," said El-Kalak.

The two men continued to watch the great white bear on the ice floe. Suddenly a seal broke the surface of the sea, and the bear struck. He dragged the struggling seal on to the floe and with one cuff crushed its skull. It was then that El-Kalak fired. The bear fell dead next to its prey. One bullet: one bear and one seal. The Eskimo smiled.

One of the most significant facts about the polar bear and its attitude toward man is that the Eskimos do not live in fear of it. They respect it, but it does not frighten them. In a way this is a paradox, for the polar bear fears no living thing in the Arctic (with the possible exception of the killer whale). Nor is there any creature upon the land that can offer any real resistance once it is caught by the massive bear. Even the wolf packs rarely tackle a polar bear. Yet for centuries poorly armed Eskimos have hunted the bear for food and clothing.

The Eskimo's method of hunting the bears requires exceptional confidence and not a little courage. He will often try to decoy a passing polar bear by lying in the snow in the hope that the bear will mistake him for a seal. Then he will either spear the animal with a knife on the end of a long handle or provoke it into falling upon him, whereupon he digs his spear into the ground and aims the point for the bear's mouth and allows it to ram the spear home with its own weight.

Hunting the bear by kayak is no less risky. The hunters herd the animal into shallow waters when possible because dead polar bears tend to float just under the surface, and retrieving them in deep water is difficult. They herd the bear by slapping the water with their paddles. Today, with the widespread use of the rifle, hunting bears is much less risky and the bear's natural sluggishness makes shooting it even less hazardous. Most hunters find that the polar bear is slow to react when hit by bullets and will invariably retreat rather than charge.

Thalarctos maritimus is a monstrous animal. The great white bear can weigh more than half a ton—sometimes a great deal more. In exceptional cases it can rear up to ten or eleven feet high, which is as high as an elephant. Each of its fur-soled forefeet weighs fifty pounds, and it can crush a man's skull with a single blow. Its claws are its main weapons, for they are heavy and sharp, but its teeth can inflict terrible damage.

The bear's range coincides roughly with the limits of the northern pack ice, although polar bears have been known to chase migrating seals far to the south of the pack. Occasionally they used to reach Iceland and have been noticed off Honshu, Japan. The southernmost limit at which they have been found is Lake St. John in the province of Quebec, which is south of London in latitude.

The polar bear's numbers have sadly dwindled after years of ruthless hunting. Even today "sportsmen" are allowed to hunt them down by aircraft. Consequently, many authorities claim the bear's attitude toward man is changing and it now lives in fear of man and retreats rather than face him.

Perry, however, questions the logic of those who say that the ruthless hunter with the rifle has taught bears to respect man. If bears hunt alone then they must die alone, reasons Perry; if this is so, how do dead bears pass on their experiences?

Nevertheless, Perry himself provides good evidence to suggest that where polar bears have had the most contact with man, there they have the most fear. In southern Greenland, where the hunter has probed deeply and often, the bear avoids man and will retreat if shot at. In the north, where man has hardly explored, the polar bear is bold and uncomfortably inquisitive when it sees a man. Its curiosity will often induce it to rush up to a man and even sniff him all over, before ponderously slouching off. These northern bears fear nothing and will even accept scraps thrown to them. Logically one would expect these unsophisticated polar bears to be voracious man-eaters, for man, no matter how desperately he might defend himself, would be very easy prey and the bear regards everything else on the ice as food.

A few years ago I was watching two newly captured polar bears engage in a most bloody battle. I made some remark about what it would be like to be chewed up by one of these creatures, when my very learned zoologist companion said, "They are the only animals in the world that count man as their natural prey. They are the only animals that will actually stalk a man on sight in order to eat him." I had heard this before. Hunter Granzel Fitz, writing in *Sporting Illustrated* a few years ago, asserted, "Of all the animals the polar bear is the most certain to attack. He attacks not because he is mad at you. He simply wants to eat you." Other zoologists will take the opposite view. Bernhard Grzimek says rather rashly that the polar bear "is harmless and trusting." He is completely wrong when he says that "there is no evidence that they have eaten men they have killed nor even disturbed Eskimo graves." The truth, as usual, lies somewhere in between.

It is safe to say that polar bears are unpredictable, that any man who goes up to one unarmed is a fool, and that they do eat people infrequently, but that their usual attitude to man is one of curiosity rather than one of aggression, and that where they have had little contact with man they are noticeably more bold in man's presence.

Early works on the Arctic give one the impression that polar bears hardly had time to look up because they were so busy eating explorers and sailors. In retrospect it seems more likely that these early explorers were so stricken with awe at the size of the bear and perturbed by its advances that they tended to misinterpret its actions and shoot it "out of self-defense." This would have increased the population of irascible wounded bears, who might have later proved troublesome. Once on Mount Germania, Greenland, a sailor assumed he was being pursued by a man-eater when a polar bear chased him down a mountain. A bear can travel at twenty-five miles per hour, and the sailor realized he could not outrun the animal. He became petrified with fright. He stood stock still as the bear sniffed at his hands and face and then, hearing the voices of the man's companions, it lumbered off.

On another Greenland expedition, Dr. Borgen was charged by a polar bear, which knocked him down, seized his head in its mouth, and dragged him off. It then let go his head and gripped him by the arm and dragged him a little farther. Then it bit his hand and then it lost interest. Perry observed: "This irresolution of a polar bear when a man is within its grasp is typical." It seems to me that a bear does not find man at all appetizing unless it is desperately hungry. It will stalk him— "automatically" if one has to use a word—because it spends its life stalking dark objects on the ice or snow knowing that they will probably be seal, whale, or walrus carcasses. But once it finds the object is a scrawny man it seems to realize its mistake. Polar bears prefer blubber to flesh and will invariably strip a seal of its blubber and leave the flesh. Man is blubberless. Sally Carrighar has an interesting theory to explain why polar bears will occasionally eat up Eskimos. "One may wonder whether the bears believe these men to be seals. Since Eskimos often wear sealskin parkas, eat a great deal of seal meat, and cleanse themselves with seal oil, the bears may confuse these familiar-scented small men with the bear's customary prey."

Hunger, maintains Perry, is undoubtedly the prime cause of

aggression. Eskimos claim that man-eating occurs mainly toward the end of a severe winter and that on Southampton Island hungry bears will destroy huts and igloos to get at the women and children inside when the men are away.

Nelson mentions an incident in which two men were tending their seal nets when a polar bear came up to them. One man rolled over onto his back and drew his only weapon, a small hunting knife. He lay there not daring to breathe, and for fear of provoking the bear he dared not use his knife unless he had to. The bear sniffed him from his feet to his face, then nuzzled him around his mouth and nose. The other man then made some sounds that distracted the bear. Immediately the bear charged, killed, and consumed the second man.

Gerrit de Veer, the Dutch chronicler with one of Barents' expeditions, described a similar incident in which a polar bear killed a man and began to eat him as the man's companions tried to frighten the animal off. The bear took a swipe at the most daring would-be rescuer and smashed his skull. Then it continued eating. Expedition members hurled anything that came to hand but the polar bear merely chased each object like a playful puppy chasing a stick.

Another factor that might cause a polar bear to take to man-eating is old age. Illingworth tells a story of a radio operator named Gibbon at Resolute Bay in the Canadian north who was dragged from his shack by a polar bear, which "no doubt" intended to devour him. Gibbon rammed his fist down the bear's throat, and this action gave a colleague time to put a bullet into the bear's heart. The bear was found to have blunt claws and teeth and to be extremely thin "due to age." "It was probably forced to attack a radio operator because its teeth were incapable of penetrating tougher meat," says Illingworth. It is puzzling, though, how this polar bear knew that radio operators were more tender than seals.

One further evidence of polar bears' dislike of man's flesh is in the number of Eskimo corpses that bears tend to dig up, yet fail to eat. Bears generally, and not just polar bears, tend to dig up areas that feel hollow or soft; and just possibly the feel of the earth rather than the scent of the corpses makes the polar bear dig up graves.

The grizzly bear, the malevolent god of the American Indians and the coveted trophy of every man who called himself a hunter, once roamed almost the entire West of the United States. But "Moccasin Joe" is down to his last few hundred now and, although many men are unmoved by his decline, some men—mountain men, sportsmen, and naturalists—regret it with a pang of sorrow. Sorrow for the great shambling, morose, rooting bear—the biggest land carnivore and a relic of the wild Northwest that is now fast becoming extinct. Today, the last 750 have left the open country where they used to live. Either they have taken refuge in the thickets of the river valleys of Colorado, Wyoming, Idaho, Montana, and maybe the state of Washington, or they have taken to undignified rooting among tourists' refuse in the national parks, where they stand at least a chance of survival. In Canada and Alaska there are still pockets where the grizzly can hunt more or less unmolested but even there his days are numbered, for the Council of the Northwest Territories now allows anybody with a game license to kill grizzlies except in the three game reserves.

In North America, grizzlies are shot in self-defense; but, although the bear can be a potential man-killer in the rutting season or when in cub, it is questionable how many of those shot were really attacking. In many parts they have been exterminated purely on the grounds that they were ferocious. *Ursus horribilis* (even the scientists who named it were carried away by its reputation) has a lot to live down. Goodman described it in *American Natural History* as "dreadful and dangerous." Almost every reference book of the early days described this powerful, long-clawed bear as a man-eater which attacked without provocation.

When unarmed zoologists and naturalists went to have a closer look, they found the grizzly to be less formidable than his reputation. Buffalo Jones, the plainsman who became a warden in Yellowstone Park, once nonchalantly roped two to remove tin cans from their feet. And at the beginning of the nineteenth century Zebulon Pike wrote that the grizzlies "seldom or never attacked unprovoked, but defended themselves courageously." Halliday claims that the most one needs in grizzly country is a stick with which to make a noise on the trees and bushes "so that he will know you are coming and avoid your presence." Although the grizzly can fell a steer with one blow, Halliday insists that it will rarely kill a man outright. Once in contact with a man it seems to fumble. "It would appear," says Halliday, "as if he has discovered his mistake after attacking." This characteristic "fumbling" may have led to the story that bears hug men to death. The "bear hug" is a misnomer. What happens is that when it attacks a creature as small as a man it tends to tear at him and bite in much the same way as a lion or other large beasts of prey. To do this it usually pulls its victim against its body.

A classic incident involving the grizzly which serves to illustrate many characteristics of the bear's method of attack and its effects is found in W. Robert Foran's *A Hunter's Saga.* In 1911, Foran and a friend, whom he calls "John E———," came upon a he-bear on Pikes Peak in the Rockies. John fired but failed to kill, and then both hunters panicked when the bear charged them. Once they began running both men had committed themselves and they both apparently realized that their only chance now lay in reaching a bluff ahead of them and then turning and firing. But running from a grizzly through a forest thicket is madness, for the bear's weight enables it to crash through the undergrowth with ease. John suddenly tripped and fell. The bear, which until then appeared to be unsure which man to pursue (for the men were fifty yards apart), immediately bore down on John. Nine feet from him it reared up on its hind legs, John shouted for Foran to fire and at the same time he emptied his rifle into the bear's chest. Foran was helpless for he was in danger of shooting his companion if he fired at fifty yards. The bear knocked John's rifle flying and then began to hug him, at the same time ripping at his body. It rolled over with its victim until the hunter lay beneath it. The hunter struggled desperately and managed to drag out his hunting knife, with which he stabbed the bear several times. Then Foran saw the knife drop from John's hand, and he knew the fight had ended. Foran risked a shot and killed the bear. He managed to find their Indian guide, who had gone off to examine his traps, and the two men pulled the bear off the hunter, who was still alive although gravely mauled. The man's breathing was barely discernible, and Foran and the guide decided to move their camp to where the man lay instead of moving him. For days his life hung in the balance before he began to improve. In three weeks they were able to move him to the hospital, where he remained for several weeks. He was horribly disfigured, according to Foran, and it was little wonder that he decided to retreat from civilization and live among the mountain men in the Rockies. There, many years later, John was killed in yet another hand-to-hand encounter with a grizzly.

Sally Carrighar notes how grizzlies appear to attack hunters without discernible provocation and puts this down to the fact that they can smell their guns. She claims that men with guns arouse the fury of grizzlies, suggesting that if they had been unarmed that would have been safer. This is not as fanciful as it seems, for there are a number of hunters in America and Africa who strongly suspect that some animals can smell a rifle and associate it with danger. And why not? A well-kept gun has a most distinctive oily smell which must carry a great distance.

In the last century, when there were more grizzlies around and when man and the bear really began to wage their final territorial war, a number of grizzlies turned rogue but few of them lasted long. Many bears turned rogue when the pain from wounds drove them mad. Others became rogues as their hunting grounds were invaded by men and cattle; they then acquired a taste for domestic stock and gradually lost their great fear of man. One of the worst rogues in the colorful history of the Northwest was Old Mose, who died in the United States in

1904 after having killed five men and eight hundred head of cattle.

As a man-eater the grizzly hardly rates. As far as the Canadian grizzlies are concerned, David A. Munro, director of the Canadian Wild Life Service, tells me there is not a single man-eating incident on record. In the United States one of the handful of man-eating cases involved a hunter named Sam Adams, who disappeared in October 1958 while hunting in the Missoula region of Montana, near the Canadian border. A search could not be made until the snows cleared some months later. Then searchers claimed to have discovered his smashed rifle, his wallet, and his shoes, which bore the tooth-marks of a bear. They also claimed to have found bear droppings containing human hair, bones, and cloth.

The grizzly is a man-killer rather than a man-eater and a man-mauler rather than a man-killer. In 1959 and 1960 there were two quite serious cases of mauling in Banff, Alberta, and Ottawa. The Alberta incident appears to have been triggered off when a campfire suddenly flared up and a bear charged two campers, biting and scratching them. In neither incident was the attack fatal.

In spite of all this the grizzly has little to answer for. For an animal that is so sullen, strong, and fast, and for one that is so mercilessly harassed in so many areas, it does surprisingly little damage.

Theodore Roosevelt described how he was once charged by a grizzly which he had shot just below the heart. Roosevelt managed to drop the bear with his second shot. He commented: "This is the only incident in which I have been regularly charged by a grizzly. On the whole, the danger of hunting these great bears has been much exaggerated."

America has three main types of bear, the grizzly, the polar, and the black, and the black is a veritable pygmy in comparison with the other two giants. In fact, apart from belonging to the family Ursidae, it has hardly anything in common with the other North American bears. It is by far the most numerous (there are an estimated 3300 in Pensylvania, 1100 in Virginia, and not many fewer in the state of New York) and, unmolested, it is a gentle animal. The few attacks by this two-to-three-hundred-pound bear which are on record were not, so far as I can ascertain, fatal or even very serious and most were caused by captive bears. The black bear, *Ursus americanus,* is normally extremely shy and in game reserves where, after years of contact with tourists, it has lost this timidity it can become a nuisance. The *Encyclopaedia Britannica* puts it in a nutshell: "It appears to be tame but in fact remains a powerful and potentially dangerous wild animal."

The brown bear, scattered in small pockets through central Europe and across Asia to Japan, is potentially dangerous but can hardly be rated a man-killer and certainly not a man-eater. *Ursus arctos* is closely related to the grizzly but is much smaller and more even-tempered. Nevertheless, like all bears, it is unpredictable.

There are a few hundred brown bears left in Europe—in Spain, on the French side of the Pyrenees, and in Yugoslavia; there is also the odd straggler in Austria and Germany. In Europe the only fatality in recent years (outside a circus or a zoo) occurred in the Djurakovac district of southern Yugoslavia when a peasant woman was killed and seven people injured after a bear went berserk in a maize field. The incident was triggered by Selmaj Beganaj, a peasant, who with his wife tried to shoot a bear out of their crop. The wife was killed on the spot.

Generally speaking the Eurasian brown bear (also known as the black bear) is a vegetarian of secretive habits and, when attacked by man, is normally more likely to run for its life. It weighs around three hundred pounds.

From time to time brown bears were used in Roman arena events. Among the emperors who enjoyed watching them being slaughtered en masse was Valentinian I (A.D. 321–A.D. 375), who kept two as watchdogs. He is said to have fed this pair on live slaves. Gordian I had a thousand brown bears turned into the arena one day in A.D. 237, and Nero, whenever he was short of elephants and leopards, would put bears on the program and have unarmed men fight them. One would assume from this that in Roman times the brown bears were savage; otherwise how did they qualify for the arena events? It is likely that the Romans trained them to kill and eat men by starving them and feeding them on slaves. Just before the arena event they would probably be starved again.

The Himalayan black bear, *Selenarctos thibetanus*, is a six-foot bear with a white V-shaped blaze across its upper chest. Like all bears it can be dangerous but is shy and generally afraid of man.

The bear which probably kills the most people in the world is the irritable Indian sloth bear, known as *Melursus ursinus*, which is found in the jungles of India and Ceylon. Pollock claims this small bear kills more people in Assam than does the tiger, but this is hardly credible.

Like all the other bears the Indian sloth is shy and generally easily shooed away. He is perhaps the noisiest of the bears when foraging for food, and his favorite delicacy is termites. Occasionally and without much provocation sloths can turn nasty, and although there is not a single instance of them turning man-eater they have inflicted appalling and often fatal injuries. Typical of the bear family, the sloth will charge—surprisingly fast—and then, just before closing with its victim, will rear up. This is normally the end of it, for a shot at the bottom of the pale-colored mark on its throat drops it stone dead. Given a chance, though, the five-foot sloth, from a standing position, will slash at a man's face with its heavily armed forefeet and this is often enough to remove the cheek and even a man's nose and eye. The sloth's three-inch-long claws can disembowel a man. It uses its teeth as secondary weapons.

Once, Kenneth Anderson shot a sloth bear 105 miles north of Bangalore in the Nagvara Hills which had allegedly killed "some twelve persons." It had injured twice that number. It is unusual for any bear to become a rogue man-killer but when it does happen it usually is the sloth bear. Anderson records that half the injured had lost one or both eyes; some had lost noses and some had lost cheeks. Those who had died had all lost parts of their faces. Anderson was told that "at least three had been partially eaten" but this was never verified and although the sloth will occasionally eat carrion there is no sound record of it having eaten human flesh.

Colonel A. E. Stewart, the big-game hunter, states: "Make no mistake about it, he comes under the heading of 'dangerous game.'" Stewart says the sloth has poor sight, poor sense of smell and is a poor tree-climber and "a man up a tree is fairly safe." Stewart also maintains that villagers fear the sloth more than they do any other animal, "and after you have seen some of the maimed faces and shoulders in the various villages you will fully realize the reason of their fear; he can make a terrible mess of a man's framework."

Most victims are attacked when they walk almost slap into the bear. The bear makes such a noise when feeding that it does not appear to hear people approaching and, presumably, the victims do not hear the bear because they are either talking or making a noise while walking. "If you surprise him," says Stewart, "the chances are he will charge at once; not that he wants your carcass as food, but simply, I believe, from fright and nervousness . . . whatever you do don't run for it; you haven't got a hope."

The Wolf

"Like one, that on a lonesome road
Doth walk in fear and dread,
And having once turned round walks on,
And turns no more his head;
Because he knows, a frightful fiend
Doth close behind him tread."

Samuel Taylor Coleridge

In the early 1960s the Canadian wildlife writer Farley Mowat wrote an appeal for the protection of "the good-natured wolf." A few years later the British writer John Pollard, who investigated the history of wolves in Europe, observed that the sewing up of captured wolves' lips and then flaying them alive "may seem cruel, but the wolf is a cruel beast and could scarcely expect to be treated better."

It is one of those inexplicable things in nature that the wolf, *Canis lupus,* has never been known to eat a man in North America and yet in Siberia and down through Europe to the Iberian peninsula it has been responsible for reigns of terror that make the depredations of man-eating tigers and lions read like humor.

According to the early American writers wolves were numerous in the New World and their depredations right through the States cost the settlers dearly in terms of livestock. The position was aggravated by the wolf's natural prey being decimated, forcing the animal to raid the corrals. The unremitting war against the timber wolf and the gray wolf (they are different races of the same species) ended with the animal's almost total annihilation south of Alaska and south of the Canadian north.

In Canada, apart from an isolated attack in 1963 when a man was bitten about the neck, and an incident in Churchill,

Manitoba, in 1926, when a delinquent and probably rabid wolf staggered into the town, there are no incidents of any significance on record. The Churchill incident caused a great deal of excitement at the time and the press, in some parts, made it sound like a major disaster. All that really happened was that a wolf was seen entering the town but later disappeared. The army was called in and in a very short time half a dozen huskies and a Chippewayan Indian had been shot. The wolf was later accidentally run over by a tourist's car.

A measure of the geniality of the wolf in Canada is admirably illustrated in a delightful book called *Never Cry Wolf* by Farley Mowat. The author has crawled into a wolves' den in the Arctic after the wolves had pulled out for fresh fields. When he reached the end of the cavern he switched on his flashlight and found himself nose to nose with a she-wolf and her cubs. She cringed and Mowat ran out with magical speed.

On the other side of the Atlantic where the legend of *Le Petit Chaperon Rouge* (Little Red Riding Hood) was born, the wolf has been, and in parts still is, a menace. The popular image of a pack of wolves, eyes glowing like candles, whooping and howling as they chase after a fleeing sleigh pulled by a snorting, frothing horse, was born not of legend but of fact. The big bad wolf of which our children are frightened gave our forefathers an extremely rough time.

Before looking at this vicious side of the wolf, however, some mention should be made of two animals that are not strictly canines: the Cape hunting dog and the dhole.

The Cape or African hunting dog, *Lycaon pictus*, is a voracious dog with a method of hunting that repulses most hunters. It hunts in a pack and pulls pieces out of its victims even as it runs along. Although the dog is not a true member of the genus *Canis*, it closely resembles a true dog. It has characteristically large, round ears and conspicuous patterning of black, white, and pink.

Patterson claimed it will attack anything "man or beast," but there is no evidence to suggest that it does in fact tackle humans. *The Field,* many years ago, had some correspondence on the subject and it mentioned an incident in which a man and a woman were carrying a dead pig between them when a wild dog pulled the woman down and killed her. As the woman was covered with pig's blood at the time it seems the dog might not have recognized her as human. She was not eaten. I know of no other incidents.

The dhole, India's "devil dog" (*Cuon alpinus*) is also not classed as a true dog mainly because of a difference in dentition. The red-colored animal is a little larger than a jackal, hunts in packs, and has a reputation almost identical to that of the African hunting dog. Nothing is sacred to a pack of dholes— not even the tiger, which it tears to pieces from time to time. Perry describes them as the terrors of the jungles and mentions in passing that "in Siberia they are reported to be man-killers." Sumatrans, according to Perry, never shoot at a pack because they fear it will turn on them. I am not in a position to expand on this although it does seem possible that in hard winters in Siberia desperate packs might be reduced to hunting down anything they see, including an occasional man.

Canis lupus is probably the progenitor of most domestic breeds of dogs although it is a great deal bigger than the majority. The Arctic wolf, which is probably the biggest race, can weigh up to 170 pounds and can measure well over eight feet from its nose to tail tip. Its footprints are six inches across. Working as a pack these wolves can pull down and quickly consume any animal that might take their fancy, although, like wolves elsewhere, they are quite content to live on mice and even beetles if they have to.

In all races of *Canis lupus* the color varies considerably and might be black, brown, red, white, gray or a combination of any of these colors. Their terrific stamina is equaled only by their terrific appetite: an adult wolf can eat thirty pounds of meat at a sitting. Under normal circumstances the wolf in Eurasia is as shy and as wary as the wolf in North America, but once it starts preying upon man it develops a fascinating degree of cunning—a cunning which is legendary and which prompted the Lapp proverb, "A wolf is as strong as one man but as cunning as seven."

In 1949 the government of Finland ordered a mass wolf hunt by troops using sledge-mounted machine guns, antipersonnel mines, and aircraft. Lumbermen, herders, and professional hunters took part, and for days they combed the forests for the stock-killing and occasionally man-killing wolf packs. The result: only a handful of wolves were shot, but as soon as the army withdrew the wolves came out in force again. This postwar surge in the wolf population of northern and eastern Europe was due mainly to the temporary disappearance of the professional wolf hunters during the war years and secondly to the increase in numbers of the wolves' natural prey, which had also been spared the hunters' bullets. The postwar years also saw an increase in the number of man-eating wolves, for the animals in many parts had acquired a taste for human flesh after scavenging on the battlefield. On the Polish-Rumanian frontier during the winter of 1946–47 a pack of sixty wolves attacked a border post and pulled down two guards, whom they killed. They were driven off with grenades and rifle fire.

John Pollard claims that France has probably suffered more from wolves than any country in the world, and he produces some convincing if gruesome evidence to back this up.

His account of the "Beast of Gevaudan" is a classic episode in the long and bloody history of man versus wild animals. It was played out among the quiet, well-wooded valleys of Lozere on the Central Plateau where, for two years, the villagers and their children lived in a state of terror as a massive wolf took a steady toll.

The man-eater of Gevaudan began its attacks in June 1764 by rushing at a woman who was tending her cattle, but it managed only to rip her clothes before the cattle put it to flight. Then on July 3 a fourteen-year-old girl was killed and eaten. By October, ten people had died—nine of them children. Dragoons were called in and, in a series of massive drives, killed seventy-four wolves just before Christmas—but the beast's attacks continued and six more children were eaten before the year ended. The last death that year was at Mende, twenty-seven miles from the scene of the wolf's first attack. It ate three people over New Year's and three more on January 6 and 7. At times the wolf must have been depending solely upon human flesh.

Finally Louis XV heard about the wolf and ordered one of the best hunters in France—a man responsible for shooting twelve hundred wolves in Normandy alone—to go to Mende and seek out the man-killer. By the end of 1765 the hunter had shot nineteen wolves, but the man-eater was unruffled and the toll mounted to fifty people. Louis recalled the hunter and sent another. This time he sent his best hunter and saw that he was assisted by the best wolfhounds in France. Meanwhile the wolf was killing with impunity. At times he was eating a person a day. The new hunter, Antoine, shot twenty-two wolves but the killings continued. On June 19, just two years after the attacks had begun, a group of three hundred hunters gathered in the forest of La Tenazière. They were a determined group of men for they were told that the previous day, in the valley that lay before them, a girl had been eaten. A sixty-year-old farmer, Jean Chastel, who was posted at La Sogne-d'Auvert, saw a wolf break cover and immediately fired. The animal died instantly, and on cutting it open Chastel found a child's collar bone. The man-eater was dead. It had killed sixty people.

That its victims were children and old and helpless peasants is typical of the wolf. Almost never will one attack an adult man and rarely an able-bodied woman. Some are cunning enough, it appears, to wait until menfolk leave a farm before going after their families.

But the lone man-eater of Gevaudan, although not unique in France's history, was certainly unusual. The normal type of man-eater in Europe and Asia hunts in a pack and is usually driven to kill people by a severe winter when his more conventional prey is scarce. They normally return to their natural prey in spring. For the last two centuries, until the wolf was expelled from France, there were very few winters indeed when somebody was not eaten. In 1712, around a hundred people were killed by hungry wolves in the forest of Orleans, and many are the stories of woodcutters fighting wolves to the death with axes. There are many stories too of travelers walking through the woods when they suddenly became aware of being followed. They would sense rather than see the wraithlike shadow of a wolf dogging their steps. Sometimes there might be two and sometimes dozens. To run would be fatal. The only possible tactic under such circumstances is to walk without sign of panic. But once a wolf lifts its head and sends out its smooth, soulless howl, it is almost beyond human endurance not to throw caution to the wind and run.

Pollard tells of a traveler who was carrying a lantern past a wood when he saw two pairs of eyes blazing like candles among the trees. He knew what they were and he knew that if he tripped or fell the wolves would rush in and kill him. He still had some way to go to reach his cottage on the outskirts of St. Felix and therefore decided to walk backward, keeping the light on the wolves. He reached home safely.

The wolf is probably extinct now in France; certainly there have been no reliable reports of it since the end of the First World War, although cases of man-eating were recorded well within living memory. In 1900, for instance, a schoolteacher of St. Yrieix, south of Limoges, heard a sound at the classroom door just after her class had gone home. The teacher opened the door and in rushed a pack of wolves, which killed and partly ate her. The last death caused by a wolf in France was on January 26, 1914, when an eight-year-old girl of La Coquille in Dordogne was killed in some woods near her home.

The last wolves in the British Isles were hunted down in the late eighteenth century, and the last recorded case of man-eating in Britain was in Scotland in 1743, when two children were taken.

There are several areas in eastern and southern Europe where wolves still survive. Until sixty or seventy years ago wolves migrated across Europe, but today these packs are rare and their ancient routes have been distorted and shortened by being straddled by towns, railways, and cultivated land. Packs continued to roam the Balkan countries and over the last few years there have been sporadic accounts of peasants, especially children, being pulled down in the woods. From 1949 to 1955, 6400 wolves were shot in southern Yugoslavia alone. Wolves are still found in Italy and in 1950 in Abruzzi (less than two hours drive from Rome) a soldier was killed by wolves after bayonetting one of the pack to death. Six years later a postman was partly eaten in the same area. Wolves today often howl near Tivoli (a few miles east of Rome), and in 1963 residents in Rome's eastern suburbs heard the mournful howls of hungry wolves. In Spain and Portugal there are healthy wolf populations. In Spain 738 wolves were shot in 1948, and in Portugal's northeast corner, eleven children were eaten by wolves in 1945.

Further to the east wolves are quite common. Pollard estimates that there are almost as many wolves in Russia today as there were during the reign of the tsars.

In the village of Pilovo in Siberia, in December 1927, there was a strong movement toward the west by wolves seeking food. Each morning peasants would see that the fresh tracks in the snow were getting nearer and nearer their homes as the hungry wolves became bolder. Then the watchdogs began to disappear on the outskirts of the village. Their remains would be found some distance from the houses where the wolves had dragged them. Soon the wolves had killed all the watchdogs, and, with nothing now to warn them that the wolves were near, women and children began disappearing from homes on the outskirts. The weather grew worse and the wolf packs were reinforced from other areas. Food became scarcer and the desperate animals began making daylight attacks. One afternoon a forager failed to return to the village, and two young men went out to find him before night fell. They did not return. Three men then set out, and of these only one came home to tell how six frenzied wolves had pulled his two companions down and devoured them not far from the village.

The villagers then prepared for a siege by wolves. They shuttered their windows and barricaded their doors and waited in fear as the wolves' howls grew nearer. Soon the wolves were howling in the village streets and above the dreadful cacophony villagers heard the terrified neighing of horses as they were torn to pieces by the packs. Emboldened by the lack of resistance, the wolves began scratching down the more flimsy doors and rushing in on the families. Heavier wolves were able to batter down some of the stouter doors and the pack would sweep in.

The carnage lasted throughout the night and Pilovo's inhabitants might have perished to the last man but for an aircraft on border patrol. Looking down at the desolate white plain, blackened here and there by the shadows of forests heavy with snow, the pilot noticed packs of wolves converging toward the village. He flew lower and saw to his horror that wolves were in among the houses. The army was called in to rescue Pilovo, but by the time they reached it almost every family had suffered a loss.

The following winter also saw large movements of wolves across Russia and in one invasion in the Marmaras-Sziget region of the Carpathians, eleven village girls lost their lives.

Pollard also describes how a pack of thirty wolves annihilated a caravan in the Urals in 1914. They killed and ate men who fought to the last with the butts of their rifles. Then they ate the women and children as well as every head of livestock.

The Soviet Union has now embarked upon an extermination campaign. It is the government's intention to kill every wolf in the Soviet Union, including the ones in reserves. In one year recently 30,000 wolves were shot. The Russians claim that in a single year 10,000 horses, 35,000 head of livestock and 26,000 poultry were killed by wolves. In that same year 168 people were attacked by wolves—eleven of them were killed.

Today the risk of attack by rabid wolves is just as great as it ever was in parts where the wolf is still to be found; nevertheless it would be a poor excuse for exterminating the animal, for there are so many other warm-blooded creatures that are carriers of the disease—the worst being the domestic dog.

For sheer horror there is nothing quite so grim as a rabid wolf. Its victim knows, even as the animal walks unsteadily toward him, that this wolf spells death whether he inflicts even so much as a scratch. In Axel Munthe's *The Story of St. Michele* there is a stark description of six Russian peasants who were sent to Paris by the tsar because they had been bitten by a mad wolf. The tsar had heard that Louis Pasteur had a remedy for rabies. They all died and as the last three became inevitably insane the nurses at the institute fled the ward, unable to stand the horror. Pasteur realized he could do nothing and he also realized that as long as these three raving peasants were alive, they were in danger of biting other patients and thus spreading the dread affliction. The great man did the only thing he could do: he gave them drugs and put them out of their misery.

In December 1839 the villagers of Puy-de-Dome in the Massif Central heard the high pitched howl of a solitary, demented wolf echo through the valleys of the wild country beyond their village. The following day, according to Pollard's account of the story, the wolf entered the village and savaged a woman and a stonemason. Three weeks later the stonemason showed the hideous signs of rabies and died in a fit of madness. He almost bit the priest who administered the last rites. Then a forester was bitten and he died a raving lunatic.

The fourth victim was a young sawyer who managed to lock the wolf's head in the crook of his arm and he called to onlookers to get a knife and stab the animal. But nobody had the courage. The man was forced to release the animal, which inflicted only a scratch on the sawyer's ear. But it was enough— three weeks later he rushed naked through the village saying he was going to commit suicide and warning people not to come near him. But he was caught and locked up, and died in a fit.

The wolf then attacked a small boy and three women, all of whom died of rabies. It moved to Charraux, where it bit a boy whose screams brought an old widow rushing out to help him. She was bitten and so was a man who came to her rescue.

The man, a peasant named Nigon, thrust his hand down the animal's throat in an attempt to choke it but he failed. All three died.

The wolf next appeared at Les Hayes and attacked a mother and her baby. The mother was badly mauled around her forearms but the baby escaped unhurt. The mother boiled up some olive oil and plunged her arms into it. The agony was worth it, for the woman survived. The wolf traveled north, bit three more people, and doubled back west, where it attacked two young sisters, one of whom fought it with a pitchfork. It then returned to Puy-de-Dome and attacked a woman and her children but miraculously they all survived, for the wolf vented its spleen mainly on their clothes. Then the wolf disappeared and is supposed to have died. It had killed seventeen people.

The general practice in earlier days was to kill the victims of rabid wolves as soon as they showed symptoms. The first sign of rabies is usually a very strong reaction to water caused by the person's desperate need for something to slake his thirst and by his total inability to drink it. The very sight of water is thought to create such violent and maddening constrictions in the throat that it sends the victim mad with frustration and pain.

The call to exterminate the wolf on the grounds that he is a cattle-killer is a little hasty and one hopes that the Russians, if they do carry out their intentions, will at least spare wolves inside the reserves. The wolf may be guilty of some dreadful crimes but in certain areas skillful game management and other conservation measures can help create territories where the wolf can become "rehabilitated."

The Hyena

". . . full of dead men's bones."
Matthew 23:27

The African's face ended below his cheekbones: his nose, palate, upper teeth, tongue, and almost his entire lower jaw were gone. Only his eyes and the upper part of his head remained intact and yet he was alive and moderately healthy and had taught himself to swallow food. He had received one bite, just one snap, his friends explained, and that was all there was to it.

The hyena had come during the night as they always do and had smelled the aroma of food around the village. It had gone deep into the village until it reached a hut, where the smell of food was strongest. There it had stopped and then must have seen the men sleeping outside the hut, for the night was warm. It loped forward to the nearest man and again paused indecisively six feet from him. It then moved silently forward until its nose was an inch from the man's face. The sleeping man, gradually becoming aware of the appalling odor of the animal, opened his eyes. But there was little he could do. The animal clamped its terrible jaws over his mouth and nose, and with the ease of a man eating a biscuit, it bit cleanly through.

Hundreds of Africans throughout Africa have had their faces mutilated or their limbs severed by the spotted hyena (*Crocuta crocuta*); a few Europeans too, hunters in the main, have been badly injured. I have seen a white woman who was so badly disfigured after being bitten in the face while sleeping in a tent that for months after the incident she had to wear a cloth mask. Apart from their dreadful hit-and-run methods the hyenas are also responsible for an annual toll of lives caused either by outright man-eating, for which the hyena is notorious in certain parts, or by victims dying from infected bites or through loss of blood.

Because some tribes of central and east Africa leave their dead—and even their dying—in the bush for predators, a number of animals grow used to eating human meat. The reason they leave their dead in the bush comes mainly from superstition but partly as a practical way of disposing of them: the scavengers of the wilds are very thorough. Especially efficient is the hyena, which eats enormous numbers of dead men. These animals will even dig up recent graves to get at the bodies, which they must be able to smell. This is the reason that many Africans cover graves with rocks and thorns.

So far as habits are concerned, there is a great deal of difference between the type of hyena that bites off a man's face or limb and the one that kills and eats its victims. The former is probably behaving normally. In Africa both *Crocuta crocuta* and the striped hyena, *Hyaena hyaena,* have the habit of foraging around camp sites and on the outskirts of villages. Because of this they grow used to people although they remain wary of them. They are nonaggressive and, except in freak circumstances, never attack a man who is aware. Hyenas are essentially eaters of dead meat and only in certain localities where there might be a lack of dead meat will they take to killing their own prey. Frequently hyenas enter camps at night, drawn irresistibly by the smell of meat and skins. They might steal and eat anything from a saddle or a pair of shoes to a tanned skin or a rifle stock —if it smells of blood. They even carry off cooking pots. I have heard of them snatching canned food and crunching open the tins in their jaws. The strength of their jaws is fascinating. They can leave tooth marks in forged steel and many an African tribesman has had his hunting spear chewed until it was unusable merely because he forgot to clean the blood off the blade.

Few four-footed creatures are quite so repulsive as the spotted hyena and certainly none is so despised. One has to hear them

eating the bones of some carcass in the veld or hear them snig-
gering, giggling, moaning, shrieking, and screaming with "laugh-
ter" to appreciate the horror that they can engender. They are
enormously strong animals with large heads and black, powerful
jaws. I once saw a demonstration of the strength of a hyena in
Zululand. We had shot an impala at night and, in response to
the shot, the hyenas had come rushing in from some distance
away to see what was dead. We split the buck carcass into two,
down the middle, and left half in the veld in the lights of our
vehicle. An adult hyena loped up—he must have weighed 120
pounds—and without checking his stride he picked up the forty-
pound half-carcass in his jaws and trotted away, holding it clear
off the ground. Occasionally hyenas of both species will bite at
the undersides of live cattle and then eat their entrails as they
fall to the ground; they will also patrol around the fence of
a cattle kraal and snatch off the muzzles of any cattle that put
their heads through the fence.

In the Mlanje district of Malawi, beginning in September
1955, there was a particularly bad period of man-eating. The
village idiot was killed and devoured on a well-beaten track
between two villages. Fred Balestra, a local farmer, who later
shot a pair of man-eaters in this district, was unable to discover
if the man had been asleep at the time of the attack, but he
believes the killing was done by one hyena. All that remained
of the victim were a few shreds of clothing and a patch of blood.
It appeared that the killer shared his meal with four or five
others.

Although this particular killing was the first of a chain of
killings which went on for several years, the peasants to whom
Balestra spoke appeared to know all about man-eating hyenas.
They referred to them as *lipwereri* and said that they were bigger
and stronger than the normal hyena, which they called *fisi*. The
first hyenas shot by Balestra were 158 and 170 pounds, compared with the average hyena weight of 128 pounds.

Seven days after the idiot was killed, an old woman was
dragged screaming from her hut one night by a hyena which
had broken through the straw door. The hyena dropped her when

a man rushed up, but by then her arm was missing and she was badly bitten around the neck. She died the next day. The third and last killing that spring was a child of six, who was killed while sleeping on a verandah. The hyena had killed her by biting her face off and then, along with other hyenas, had eaten her entire body, leaving only the head. Balestra feels that they left the head only because they were disturbed by village dogs. The following year five people were taken and from then until 1962 the number fluctuated between five and eight each year. Seventy percent were children. In 1961 all six victims were children—four were eaten in January. All the attacks took place during the summer months, when Africans sleep outside their huts. At one stage the government used an air force plane to bomb the caves in which it was suspected the hyenas were sleeping during the day, but the raid produced little effect. The killings continued over an area fifty miles deep and sixteen miles wide.

Balestra, who was the only man prepared to hunt the hyenas, believed for a time he was up against lycanthropy, for quite often victims would be taken from the midst of their sleeping families without a sound being heard. In reality, the hyenas probably grabbed them by their faces or killed them before they were really awake. Fears of witchcraft made some local inhabitants reluctant to give information, and one can be sure that the witch doctors were encouraging these fears. This kind of superstition exists throughout central and east Africa, and there are tribesmen who sincerely believe there are no wild hyenas in the world—that they all belong to witch doctors. The belief is furthered by the witch doctors themselves, who often keep hyenas as pets. Some even train them to eat men and hire them out to settle scores. George Rushby, who worked as a game ranger in central Africa for some years, shot hyenas with beads interwoven in their hair and some of them had mysterious patterns cut into them. He also records having shot one which was wearing khaki shorts. One hyena shot by him was claimed by a female witch doctor as her lover. She gathered it up and carried the reeking beast away.

In Africa it is almost certain that the hyena is a more consistent man-eater than the leopard. In parts it is a greater man-killer than both the lion and the leopard together.

It is odd how *Hyaena hyaena,* the long-haired, pointed eared, slim, striped hyena which one finds from central Africa through north Africa, into Asia Minor and right across to India, is so similar in habits yet so different in appearance from the spotted hyena. The spotted variety is confined to Africa. The striped hycna is nowhere as common in Africa as the spotted one, but beyond the Ethiopian region the striped takes over. A report in January 1967 claimed that the striped hyena ate five children in the long-suffering Uttar Pradesh.

IV
KILLER CATS

John Pitts

The Tiger

The Lion

"Save me from the lion's mouth."
The Book of Common Prayer

The sight of a charging lion is disconcerting indeed; if most big-game hunters were a little more honest about their feelings they would describe it as terrifying. Try to imagine the heavy, shaggy-headed cat standing with tufted tail erect and its mane bristling. Slowly it begins to gallop forward, using a gait very similar to a large dog. As it draws near it contrives to flatten itself against the ground and for some odd reason appears to shrink in size. Then comes the moment when it skids to a halt, gathers itself in a split second, and leaps forward and upward, teeth bared and massive claws out.

A healthy lion that has become a man-eater would need a minimum of fifty victims a year to stay alive, and a hundred and fifty to remain in peak condition. In Malawi one was known to have eaten fourteen people in a month.

C. J. P. Ionides mentions one that ate a fat woman and then a warthog "for dessert." Ionides, who shot forty lions in his day (of which more than half were man-eaters), shot this one as it "slept it off." Individual lions have killed and eaten as many as three people in a single night. For a person who has never seen a wild lion it is difficult to appreciate how enormously powerful they are. For a creature that is capable of killing a buffalo, a man is small fry. Lions have been known to carry victims for up to a mile without rest. One judiciously placed bite—usually in the head or neck—kills instantly and so does a swipe of its great pad. One blow can break a man's neck; it can break an ox's neck.

But the fear that most people have for lions is not shared by those who really know them. Game rangers rarely go about armed in African game reserves unless they are on control work, and usually a threatening gesture or a shout is enough to send the lion loping off into the bush.

The former warden of Kruger National Park, Lieutenant Colonel J. Stevenson Hamilton, once told me at Skukuza: "Lions can be very troublesome animals, especially at night when they jump over my fence looking for livestock. They wreak havoc in my cabbage patch. I have to go there and wave my hat at them and say 'Shoo!' That scares them! They're not very courageous animals."

Lion hunting as a sport is not considered particularly dangerous by white hunters in Africa. Even in the days of the early Cape settlers in the late seventeenth century it was not considered unduly hazardous. Today some descendants of these settlers consider it unsporting to shoot a charging lion; they wait until it stops and gathers itself for the final spring and then fire into its brain from a distance of a few feet. Tobi Rochat of Acornhoek in the Transvaal once told me that out of three hundred lions he had shot, not one had given him a bad moment. His wife, a pleasant, motherly woman, has also shot a few.

Harold Hill made an interesting remark (recorded by Dan Mannix) when he heard George Grey had died after a dreadful mauling by a wounded lion: "He should never have struggled with the lion. That only angers the cat. I remember once when a leopard jumped me . . . I fell on my face and gave him my arm to chew. After a while he got tired of it and left me."

João Augusto Silva, the Mozambique hunter, claims that the female will abandon her cubs in the face of danger rather than fight. This may be so—leopards are also like that—but hunters find that the lioness will invariably charge when her mate is fired at or wounded. (Male lions are not anywhere near as chivalrous.) Richard Perry states, "Lions have always been much

easier to kill than tigers, being bolder and less cunning, asso-
ciating in tribes and inhabiting more open country." Elizabeth
Balneaves writes about a man named Brooks who killed two
lions with one bullet: it passed through the jugular of one,
killing it instantly, and punctured the jugular of one behind it,
killing that one instantly.

On the other hand, some hunters, such as Denis D. Lyell,
claim lions are extremely dangerous. "There are more fatal
accidents from lions on record than from any other animal," he
says. I personally believe that the elephant has killed more
people in "accidents" than the lion, but the margin, if indeed
there is one, is small. Crocodiles and snakes kill more humans,
of course, but not necessarily in "accidents" where the human has
been guilty of misjudgment. The lion is second only to the
crocodile as a man-eater in Africa.

A "normal" lion, one that is not a man-eater, is not a great
killer of humans at all. It has an obvious fear of man and
will usually slink away as soon as it catches man's scent. I have
seen a lion bolting on only two occasions: once when charged
by a bull elephant and once when a man, racing through some
bushes beside the Olifants River in South Africa, ran slap into
one. Who was the more frightened is difficult to say, but certainly the man regained his composure first.

Lions have probably held man in respect since the dawn
man with his crude weapons and great cunning proved himself
a superior animal. Austin Roberts writes that even today, in some
parts of Africa, tribesmen will chase lions from their kill (once
they feel that the lions have almost satisfied their hunger) and
take away the rest of the meat. Clive Cowley, writing on the
River Bushmen of the Okavango Swamps, says that the little,
pygmy-size people will also rush at gorged lions and rob them
of their prey. The lions rarely retaliate. Their respect for man
must have greatly increased over the last four hundred years,
when the firearm has become commonplace on the African
continent.

What of hunters who are killed by lions? From my own
records it seems 75 percent are killed by wounded or sorely

provoked lions. The great majority of these attacks occur in dense bush to which hard-pressed or wounded lions have retreated. A lion that has succeeded in killing a hunter, unless it is a man-eater, will almost never show an interest in eating its victim. A great many men who survive the initial mauling die of infection days or even weeks afterward, infection from the decaying meat lodged in the lion's claws and teeth. Game warden Bruneau de Laborice of French Equatorial Africa died weeks after receiving a bite on the arm from a lion and there are several similar incidents. In recent years the use of penicillin has undoubtedly saved hundreds of lives.

The now illegal Masai ceremony of spearing a lion is well known. Last century the Masai provoked a lion into charging a youth who was then supposed to demonstrate his worth in order to be classed as a *moran* (warrier). As the lion sprang, the youth would plant his spear, point uppermost, firmly into the ground and guide it so that the lion impaled itself. Later the ceremony became one of communally spearing the lion and the man who first pulled its tail would get the honors. Following the ceremony, which is very rarely practiced today, several men would occasionally die from wounds and infection.

According to some, the charge of a lion is easily checked—even without a gun. Heinrich Lichtenstein, the German zoologist who lived for some time in the Cape, quotes local farmers as saying that an unarmed man was in no danger if he stared a charging lion in the eye "with perfect steadiness and composure. . . . the lion is then supposed to become confused, retreating slowly at first and then faster and faster. . . . Hasty flight by the pursued person, on the other hand, was said invariably to induce the lion to take up the chase, the unfortunate person soon being overtaken and killed." Lichtenstein had reservations about the advice: "I cannot help feeling that the experiment has not very often been made." Charles A. W. Guggisberg, author of a highly respected work on lions (*Simba*), comments on Lichtenstein's remarks. "The settlers' advice was quite well founded . . . although standing absolutely motionless may not be a complete guarantee of safety, it is certainly much better than running away. On more than one occasion a charging lion has turned aside only a short distance from the hunter who did not move, to chase after one of his companions who had lost his head and run away." One man who tried it was George Rushby; the lion pulled up at fifteen paces, turned tail, and fled. J. H. Patterson, author of *The Maneaters of Tsavo*, also observed that a charging lion would stop and slink away if its intended victim stood still and stared.

Once, when a lion came loping toward an acquaintance of mine, he, being unarmed, took it into his head to run at the lion. The lion pulled up, spun around, and loped off. The only sure way of ending the charge of a lion, however, is a bullet in the brain. Even a heart shot can be inadequate, for it does not always stop the animal from reaching the hunter before it dies, and one swipe of a lion's paw is enough to kill a man.

John F. Burger wrote that the lion is a particularly difficult target to hit "when coming in that peculiar gait it adopts when charging . . . surprisingly the lion's course in a long-distance charge can be changed with reasonable certainty by rapid gunfire—even from a light calibre rifle. It is when it charges from zero range that it is so deadly dangerous." Burger recalls that 75 percent of the lion victims in Nairobi cemetery died in surprise attacks from "zero range."

Leo leo (or *Panthera leo*), the second largest member of the family *Felidae*, can, in exceptional cases, weigh up to a quarter of a ton (553 pounds, Austin Roberts), can measure nearly ten feet from nose to tail tip and towers seven feet when it rises at the end of a charge. According to Guggisberg it has the strength of ten men and has been seen to clear a twelve-foot chasm when driven by desperation, and to leap thirty-six feet. Lions *can* climb trees, although they rarely do. At Johannesburg Zoo in 1966 this was magnificently demonstrated by a lioness which climbed forty feet up a smooth wooden pole.

From area to area, lions differ in color and in the size of the mane of males. Some males have no manes, some have black ones, some have pale ones. The lions of India (in the Gir Forest) are identical to those in Africa except for a difference in the muzzle, which is 10 percent broader. The Indian lion is very rare today—there are only about 250—and by nature retiring and to my knowledge has no record of man-eating.

In Africa, lions are extinct in the far south (Cape Province and the Orange Free State) as well as in the far north, where the Barbary lion was exterminated during the desert war of 1941 and 1942. From the Sudan down to the Transvaal they are common and even prolific in parts. Throughout their entire range, man-eating incidents occur infrequently, although there are districts where the incidence is curiously high. Guggisberg says of man-eating by lions, "I came across records from practically all parts of the lion's African range of distribution— from South Africa, from Bechuanaland and the two Rhodesias, from Portuguese East Africa and Angola, Tanganyika, Uganda, Kenya and Somaliland, from the Sudan and the Belgian Congo, French Equatorial Africa, Ashanti, Nigeria and French West Africa." He records the most southerly attack as being just outside Cape Town, where in the eighteenth century a soldier was eaten.

But it is Central Africa that probably suffers more than any other area from the depredations of man-eating lions. The most notorious part includes the Southern Province of Tanzania (formerly Tanganyika), the north of Malawi (formerly Nyasaland), and the northern section of Mozambique. This Central African region has bred such delinquents as the man-eaters of Njombe, which ate between 1000 and 1500 people. In the Mafienge district (just outside the Southern Province) two lions killed twenty-three in about two months. In the Lindi district just above Mozambique four pairs of man-eaters patrolled the Nyangao-Massessi road in search of victims (Guggisberg). They have a special drum in these parts called *ngula mtwe* (it means "a man is eaten") and its somber beat comprises two short thumps followed by a long one. Just south of this region, over the Mozambique border, Austin Roberts and Vaughan Kirby hunted down four man-eaters which were eating twenty people a month. In some parts, villages were abandoned from time to time when man-eating got really bad and even the local style of architecture shows certain modifications to allow for persistent man-eaters which might try to rip open a hut to get at the inhabitants. One of the worst areas in Central Africa used to be the Luangwa Valley in Zambia (formerly Northern Rhodesia) and central Malawi (Nyasaland), where in recent years a single lion ate fourteen people in a month. John Taylor once shot five man-eaters in a single night (possibly a record) after an outbreak of man-eating at Nsungu on the north bank of the Zambesi. He shot three as they tried to tear down the door of a hut.

Outside Central Africa, the Tsavo area is well known as an occasional haven for man-eating lions. Here, dozens of workers were eaten during the building of the railway from the coast of East Africa through to Nairobi. Some people believe that the bout of man-eating was triggered by the quantity of dead fever victims whose bodies had been thrown into the bush to save burial. But Tsavo was notorious long before the railway was built and has been hunted by a few man-eaters since those unhappy times. In 1955 a telegram was received in Nairobi from a Tsavo Railway official who wrote, "ODEKE NARROWLY ESCAPED BEING CAUGHT BY LION . . . ALL STAFF UNWILLING TO DO NIGHT DUTY. AFFORD PROTECTION." A generation ago, Foran shot four man-eaters in a day near this area after they had killed fifty people in three months. Since the Second World War, however, there has been only one case reported. An emaciated lion (380 pounds) was shot at Darajani, not far up the railway line from Tsavo, in 1965. John Kingsley-Heath, who shot it, found a porcupine quill rammed into its nostril.

The Ankola district of Uganda, just west of Lake Victoria, has a bad reputation. The Sariga area of Ankola is particularly notorious and in the 1920s, according to Guggisberg, a band of man-eaters roamed over hundreds of square miles. One of them accounted for eighty-four victims and another ate forty-four. Their depredations spread to Entebbe on the north shore of Victoria. This particular reign of terror ended after seventeen lions had been shot. But in 1938 another began. Guggisberg attributed the outbreak to rinderpest, which had killed off the game, forcing the lions to eat cattle. Rushby states that rinderpest killed off game and cattle, and the human population starved. The human dead were left in the bush and in this way the lions developed a taste for human meat. Within a year of this outbreak, there were eighty dead and more than half a dozen man-eaters on the prowl.

What makes one area attractive to man-eaters and others not? I suspect that in Central Africa one cause was the Arab slave trade. Over three centuries, Arab slave traders left hundreds of thousands of unwanted captives to die near their villages and thousands more beside the well-beaten slave trails that cut through the hostile African bush. Practically all of them were cleaned up by lions, leopards, and hyenas; and, as the Arabs were taking the cream of the African tribes, the useless few whom they left behind were powerless to keep back the wild animals. Man-eating became habitual among the big cats and there is still evidence of this today. John Taylor observed that on the Revugwi River, near its confluence with the Zambesi, one could shoot out all man-eaters, but sooner or later they would reappear.

Several other factors also contribute. When the monsoons come the grass grows too high for the lion to hunt its normal prey, and during this period some appear to turn man-eater. The disappearance of game through hunting, snaring, and veld fires may also force lions to hunt either livestock or humans. The custom some tribes have of leaving their dead, and even their dying, outside the villages to be eaten by the carrion eaters does not help either. Sometimes the lion gets there first and acquires a taste for people. Man-eating frequently flares up after human catastrophes such as wars or epidemics when, presumably, dead or dying people are in abundant supply.

It is significant that a great many hunters describe man-eaters as being in good condition. The Njombe man-eaters and the Tsavo man-eaters were healthy specimens. It tends to disprove the theory that most man-eaters are forced to eat people because they are either too old or too crippled to hunt game. A lot of sick lions do turn man-eater, but sickness is not as important a cause of man-eating as many people believe. (Old lions, however, are capable of taking up the oddest diets; they catch mice, frogs, lizards, and guinea fowl and even follow hyenas around in the hope of snatching carrion.) A number of man-eaters have never hunted anything else. They were born to man-eating parents, who weaned them on human flesh and taught them the art of stalking men. These can be the most difficult to hunt. Some are never caught and die of old age.

It is difficult to rank the lion among the man-eaters of the world. Lions usually hunt in tribes, and so it is impossible to establish what any individual lion ate. From my examination of the records, I would put the lion in fourth place today, after the crocodile, shark, and tiger.

Man-eating lions display a number of idiosyncrasies that are not found in "normal" lions. One of the most pronounced is their habit of moving as far away from the kill as possible before dawn. They might cover twenty miles after eating a victim. A man-eater can range over hundreds of square miles. It is hard to tell whether it does this knowing that this movement makes it almost impossible to track, having struck at a village once, it knows there is little chance of catching another victim so easily. Another idiosyncrasy is the man-eater's reluctance to eat anything but human flesh and his enormous patience in seeking it. Perhaps the most odd characteristic of all is the extraordinary lengths to which lions go to get a particular victim. It will step right over a sleeping man in order to get at one beyond him. C. J. P. Ionides wonders whether it is a difference in each man's smell. Quite often, once a lion has set its sights upon a particular victim, it will attempt to tear down doors to get at him, and burning torches will not deter it. In Kenya, according to Ionides, a man-eater entered a ring of blazing fires to snatch up a sleeping man. Taylor describes how a man on a bicycle rode down a hill straight in between two man-eaters. The lions pulled the next cyclist down and ate him. Guggisberg gives another example: An African near Fort Mangoche, Malawi, was attacked outside his hut, which was isolated from the village. His wife rushed at the lion with a firebrand and the lion dropped the man. The woman dragged her husband inside and bolted the door. He died a few minutes later. Meanwhile the lion was attacking the door trying to tear it down. The woman picked up a firebrand once more and rushed out into the night—the lion then re-entered the hut and carried off its victim.

Ionides describes how stealthily a lion can select a victim. Heinedi Ngoe was given a break from hunting a man-killer in Tanzania. The lion was blamed for the deaths of forty-three people, and the hunt for it had been long and arduous. Ngoe, who decided to take advantage of his break and get married, became victim number forty-four, and although he was snatched as he lay beside his sleeping bride nobody heard a thing. The alarm was raised by the bride, who woke up during the night to find her husband no longer with her. As she felt for him she discovered his pillow was blood-soaked. She screamed and villagers had to restrain her from going into the bush to find Ngoe. Next morning Ionides found that the lion had waited just outside the village for a considerable time with its prey before going deeper into the bush.

The tenacity and cunning of a man-eating lion are again illustrated in an incident in June 1900 at Kima Station in the Congo, west of Africa's Great Lakes system. The lion attempted to tear through the corrugated iron roof of a station building and eventually succeeded in carrying off a railway driver, but the man managed to squeeze into a galvanized iron tank. For several minutes the lion tried to hook him out by placing its paw through a hole like a bear at a honey pot. Superintendent Ryall of the Railway Police decided to have a try at shooting the beast and had his personal railway carriage shunted into the siding. Two others, Hubner and Parenti, joined him. The three men occupied one compartment, and Ryall took first watch by sitting on a bunk in the compartment and keeping watch out of the open window. The two others slept, Parenti on the floor and Hubner on a high bunk. Ryall must have nodded off to sleep. The lion, through either luck or incredible intelligence, entered the carriage undetected at one end, padded softly down the corridor and slid the door open. Hubner woke up and saw the lion below him straddling Parenti, who was lying motionless but awake. The lion reached across the compartment, swiped Ryall across the side of the head (probably killing him instantly) and then sunk its fangs into his chest near the left armpit. Hubner leaped over the lion's back and tried to escape through the sliding door leading into the corridor, but frantic tribesmen, knowing what was inside, were keeping it shut. Suddenly there was a crash, and Hubner looked over his shoulder to see the lion leap through the window with Ryall in its jaws. Parenti, his Latin excitement getting the better of him, also leaped out of the window. The lion had been standing on him. The maneater was later caught in a trap and was put on show until it was shot a day or so later.

It is most unusual to find a successful man-eater stupid enough to get itself caught in a trap. It is vital when hunting a man-eater to kill him quickly; if the first few attempts to kill him are abortive, then the lion will most likely have learned some valuable lessons in evading hunters. Successive failures do worse than that; they tend to encourage local tribesmen to believe even more fervently than before that the man-eaters are under the orders of the witch doctors and are immune to bullets. These beliefs are a very real handicap to the authorities and lead local leople to refuse to cooperate, in case the bewitched lions come for them. The tribesmen will even go to the extent of erasing the spoor so that hunters cannot follow. They feel that by doing the man-eater this favor they will earn immunity.

No book gives as vivid a description of the terror created by man-eating lions as does Lieutenant Colonel J. H. Patterson's *The Man-Eaters of Tsavo* (1907), a classic in the field of big-game hunting. Patterson's account of a typical raid also serves to show the absolute fearlessness of some man-eaters. ". . . the two brutes made a most ferocious attack on the largest camp in the section, which for safety's sake was situated within a stone's throw of Tsavo Station and close to a Permanent Way

Inspector's iron hut. Suddenly, in the dead of night, the two man-eaters burst in among the terrified workmen, and even from my boma, some distance away, I could plainly hear the panic-stricken shrieking of the coolies. Then followed cries of 'They've taken him; they've taken him,' as the brutes carried off their unfortunate victim and began their horrible feast close beside the camp. The Inspector, Mr. Patterson, fired over fifty shots in the direction in which he heard the lions, but they were not to be frightened and calmly lay there until their meal was finished."

In the morning Patterson and some others set off to track the lions. Dalgairns believed he had wounded one and pointed to a dragging trail which could have been a lion's foot. "After some careful stalking, we suddenly found ourselves in the vicinity of the lions, and were greeted with ominous growlings. Cautiously advancing and pushing the bushes aside, we saw in the gloom what we at first took to be a lion cub; closer inspection showed it to be the remains of the unfortunate coolie, which the man-eaters had evidently abandoned at our approach. The legs, one arm and half the body had been eaten, and it was the stiff fingers of the other arm trailing along the sand which had left the marks we had taken to be the trail of a wounded lion." The lions got away and the coolies went on strike. Work on the Tsavo Railway came to a standstill and the killings continued night after night. Not one lion had been killed.

Soon after this Patterson called in a Mr. Whitehead, who, on the night he arrived at Tsavo, was ambushed by a lion which only succeeded in clawing his back in the darkness. Whitehead's gun went off in the fracas, and the lion switched its attack to Whitehead's askari, Abdullah. *"Eh, Bwana, simba!"* was all the askari had time to say. He was eaten. The following night Patterson sat up in a machan, a very rickety one twelve feet from the ground. It was dark and the silence flooded in. "A deep long drawn sigh—sure sign of hunger—came up from the bushes, and the rustling commenced again as [the lion] cautiously advanced. In a moment or two a sudden stop, followed by an angry growl, told me that my presence had been noticed, and I began to fear that disappointment awaited me once more.

"But no; matters quickly took an unexpected turn. The hunter became the hunted; and instead of either making off or coming for the bait [a donkey] prepared for him, the lion began stealthily to stalk *me!* For about two hours he horrified me by slowly creeping round and round my crazy structure, gradually edging his way nearer and nearer. Every moment I expected him to rush it . . . I began to feel distinctly 'creepy' and heartily repented my folly in having placed myself in such a dangerous position. I kept perfectly still, however, hardly daring even to blink my eyes; but the long continued strain was telling on my nerves. . . ." Patterson then felt a blow behind the head which terrified him beyond words. Lesser men would have, at this stage, died of a heart attack or climbed higher into the tree and yelled for help. But Patterson soon realized it was nothing more harmful than an owl. His involuntary start caused the lion to growl. It now began to advance. "I could barely make out his form . . . I took careful aim and pulled the trigger. The sound of the shot was at once followed by a most terrific roar, and then I could hear him leaping about in all directions." The lion bounded off but was not far away when Patterson heard it plunging about. He sent more shots after it and then heard it groan. Gradually the moans became long, deep sighs and finally these stopped. The first man-eater of Tsavo was dead, and, as Patterson shouted the news from his machan, hundreds of coolies in nearby camps yelled, *"Mabarak! Mabarak!"* (Savior). The lion was an excellent specimen, measuring nine feet eight inches from nose to tail tip and three feet nine inches high at the shoulder.

R. O. Preston was also on the construction of the Tsavo railway line and one of his stories, told to Dan Mannix (*African Bush Adventures*), is an example of the trials and tribulations of the white pioneers in East Africa. A man named O'Harra, who was in charge of road construction along the Tsavo Railway, was seized in his isolated tent by a man-eating lion, but Mrs. O'Harra's screams and the man's struggles confused the lion and it ran off. A few minutes later the cat stole back to the tent, but by this time Mrs. O'Harra had the gun out and she kept the animal at bay until help arrived next morning. O'Harra was so badly mauled that he died, and during the whole grim episode the O'Harras' children cowered in a corner of the tent.

Colonel Patterson gives a vivid account of the strength of the lion in his Tsavo saga:

"The lion managed to get its head in below the canvas, seized him by the foot and pulled him out [of the tent]. In desperation the unfortunate water-carrier clutched hold of a heavy box in a vain attempt to prevent himself being carried off, and dragged it with him until he was forced to let go by its being stopped by the side of the tent. He then caught hold of a tent rope, and clung tightly to it until it broke. As soon as the lion managed to get him clear of the tent, he sprang at his throat and after a few vicious shakes the poor bhisti's agonizing cries were silenced forever. The brute then seized him in his mouth, like a huge cat with a mouse, and ran up and down the boma looking for a weak spot to break through. This he presently found and plunged into, dragging his victim with him and leaving shreds of torn cloth and flesh as ghastly evidence of his passages through the thorns."

This lion showed complete disdain for fire and human shouts, for it ate its meal in the light of the campfires, leaving only the skull, jaws, a few large bones, and two fingers.

Man-eaters leave little behind unless they are really harried. When the victim is a child, the lion eats all except the skull-cap, which is licked clean. All other bones are usually eaten. In the case of adults, the lion usually leaves the skull and jaws (because of the teeth), the soles of the feet or boots (often with the feet still in them) and bones such as femurs and the hips. The meat is licked off these larger bones by the cat's rasplike tongue.

Guggisberg points out that prompt action can sometimes save a man-eater's prey. The victim might remain alive for some time after being carried off. There was a case involving a man in German East Africa who called—in vain—for fifteen minutes while literally being eaten alive.

Hans Blesser also tells how he stood by helplessly when a villager was dragged screaming into the bush by a man-eater. Blesser got volunteers together and they advanced on the lion who, with determined charges, drove them back time and time again. All the time, says Blesser, the victim moaned for help or screamed when the harried lion fed on him. Next day his abandoned body was found intact except for a thigh and calf.

It has often been said that a lion's victims feel nothing. Livingstone, who was attacked by a lion in Bechuanaland, wrote this account: ". . . growling horribly close to my ear, he shook me as a terrier does a rat. The shock produced a stupor . . . a sort of dreaminess in which there was no sense of pain or feeling of terror, though quite conscious of all that was happening. It was like what patients partially under the influence of chloroform describe, who see all the operation, but feel not the knife. . . . This singular condition was not the result of any mental process. The shake annihilated fear and allowed no sense of horror in looking around at the beast."

Guggisberg quotes the Hungarian naturalist Kittenberger, who was badly mauled in Tanzania, as saying he felt "no pain at all." Others have repeated similar sensations. On the other hand, C. Cronje Wilmot, a Nqamiland tsetse-control officer, who was mauled and wounded twenty-three times in one attack, records feeling intense pain. Arnold Weinholt of Australia and Petrus Jacobs, who were attacked at different times, recall terrible pain—"like having nine-inch nails driven into you," said Weinholt.

A great deal of man-eating on the African continent is done by prides of lions rather than solitary lions. The Njombe man-eaters numbered fifteen, and there were even more Anhole man-eaters. The Tsavo killers were eight.

The Njombe man-eaters, shot out in 1947 by George Rushby, were slightly smaller than the ordinary run of lions, and their pelts were "glossier and more luxuriant than those of lean, hard-working, game-hunting lions." The lions, referred to in some records as the "Ubena man-eaters," operated over a range of fifty miles by thirty in the vicinity of the northern tip of Lake Nyasa. They first made an appearance in 1932 and by the beginning of the war were moving about in three or four small prides. In the game area under Rushby's administration they killed ninety-six people in 1941, sixty-seven in 1942, and then another eighty-six before their reign of terror ended. During this period they were killing at a heavier rate in a second area and at a lower rate in a third. "The renowned man-eaters of Tsavo were small fry in comparison."

Rushby was transferred to Mbeya in the Southern Province of Tanzania in 1946 and almost immediately received a telegram from W. Wenban-Smith, the Njombe District Commissioner. It read: "I BEG YOU TO APPLY EARLIEST ATTENTION TO MAN-EATERS. CONDITIONS IN THIS DISTRICT PATHETIC." Along the short Njombe section of the Great North Road, seventeen road workers had been eaten. The villagers had evolved "a negative form of defense" by drawing into larger communities and abandoning the smaller villages. But the human toll did not decline and incredibly, during the fourteen years of man-eating, not a single lion had ever been shot in the Njombe area.

A typical Njombe incident occurred in Rujewa. A lion rushed into the village, bowling people over as it went, and grabbed a woman as her husband stood frozen from shock. Without apparent effort it carried her in its jaws to a group of lions waiting on the perimeter of the village. They ate the woman in a thicket. The husband, armed with an antique rifle, was one of the few brave enough to go after the lions but as the party drew near the thicket he came face to face with a lioness. It was carrying his wife's leg. He was so shaken that he could not fire and the lioness walked off.

Rushby tried every ruse imaginable to kill the lions but each one failed, including a series of cunning traps. He realized the importance of quick success, because of the already strong tribal superstitions. The problem with traps was that they had to be baited with the meat that the lions liked best—human meat. But the remains of tribesmen were scant. Even if some could be found, where could he set the traps, since the lions would wander off a dozen miles in any direction by the next day?

Rushby had his first opportunity when a messenger brought news that lions had attacked the village of Mambego, fifty miles from where he had anticipated their striking. Two villagers had been taken. Rushby, tired from three futile hunts, raced to Mambego and spoored a lioness, which he shot with four bullets. She did not necessarily need four bullets; it was just that this one "had to be dead," said Rushby. Three more man-eaters were bagged soon after, but the toll of human life did not appear to drop. Nevertheless, the four dead man-eaters encouraged tribesmen to take up the hunt and massed hunts began. These had an immediate effect upon the death toll. It rose. Tribesmen, firing wildly, killed no lions but managed to kill three of their own numbers.

By 1947 the lions' toll was falling abruptly, except in one area where the lions appeared to be making one last stand. Rushby, splitting his best helpers into pairs, combed the bush and shot a male and female. It seemed to be all over when suddenly a woman was eaten. Two more lionesses were flushed out and shot.

In all, fifteen confirmed man-eaters were shot in the Njombe district, two were injured (probably mortally) and five other lions—probably not man-eaters—had been shot. The first seventeen man-eaters had been responsible for what Rushby described as "without doubt the greatest and most sustained record of man-eating ever known in Africa." The hunt had lasted fifteen months. In as many years successive generations of killers had eaten between 1000 and 1500 people—enough to populate an entire village.

In Africa there is a firm belief among tourists and even professional hunters that lions will never attack a person in a car. The belief has grown up largely with the advent of game reserves, in which people can with comparative safety drive up to big game in their cars. In most of these parks lions tend either to ignore cars completely or to use them as a means of ambushing antelope. The car fumes mask their scent.

In March 1962 the theory received a setback when Frederick van Wyk and Ronald Holloway were attacked as they slept in their car half a mile south of the Chirundi Bridge over the Zambezi, between Kariba Dam and their home town, Lusaka. They were severely mauled as they fought the lion inside the car. They finally managed to slam the door on its body, causing it to retreat. The incident is not unique. T. Murray Smith was arriving with the Maharajah of Jodhpur in an open truck near Lake Manyara, Tanzania, when a lion sprang onto the hood and smashed the windshield. The Maharajah shot it through the brain at a range of twelve inches. Lions infrequently jump onto cars and peer at the occupants through the windshield; their motive probably, apart from the odd incident, is curiosity.

White men try to frighten themselves with such fictitious horrors as vampire bats or werewolves, but not even the top horror writers can dream up something to equal the real-life horrors suffered by Africans who are victims of the watuSimba.

Lycanthropy, to give this phenomenon its technical name, is a form of black magic which manifests itself in men taking on the form of animals. *Mbojo* is what the Swahili-speaking people call it—and the lion-men who are created by *mbojo* are called the watuSimba. It is found in various forms up and down the length and breadth of tropical Africa, but it has always been particularly prevalent in the Central Province of Tanzania.

In 1920 the police in the Singida district combed the bush and villages for the murderers of more than two hundred people. The shooting of eight man-eating lions had revealed that scores of their suspected victims had actually been killed by being stabbed through the heart and then clawed. Mr. W. Hichens, who shot the real man-eaters, described in the magazine *Wide World* how he was stalking one lion and came across a youth wearing a skin. The youth, drugged to the point of insanity, wore gloves with lion claws attached and carried long stabbing knives. People in the area where the genuine man-eaters were on the rampage were paying protection money to save themselves from the real man-eating lions, whom they assumed were being controlled by the witch doctors. John Taylor refers to the same outbreak of lycanthropy and says that there were actually eleven man-eating lions and that the witch doctors took advantage of their presence. Those in the know were able to hire the lion-men to settle old scores. The relatives of the victims would be approached and then, in turn, could hire the killers for a revenge killing. Over a period some of the lion-men were caught and hanged.

In 1946 things came to a head again when, a little farther south nearer Singida village, thirty tribesmen were murdered by watuSimba. One victim, a woman, managed to run for her life and described to the authorities how she had been attacked by a youth wearing a skin. When she had fully recovered she backed down and said she had been clawed by a real lion. In January of the following year fifty-three Africans were rounded up by the police and eight of them charged with running the lion-men ring. They appeared in Dodoma court. Because the actual lion-men were still loose and under their masters' spells, the killings continued. Before January was over, ten more people had died. There seemed little doubt that the witch doctors had also attempted to train real lions and hyenas to be man-eaters, and in 1947 a lioness, her teeth neatly filed to points, was shot in the district. In March and April a couple of dozen more people were murdered. By June the death toll of that year—in spite of the arrests—was 103. By then twenty-nine of the arrested people had been sentenced to death, but the murders did not cease until 1948. Since then there have been other outbreaks. In 1958, according to Bulpin (quoting Rushby) twenty-two lion-men victims were reported.

During the Dodoma trial a great deal of light was thrown upon lycanthropy. Some lion-men were hired out for a little more than five dollars a killing and were in turn "sublet" to others to cover the original costs. A woman testified that her husband was kidnapped by witch doctors, who drugged him and trained him to go about wearing a lion mask, with his body covered in baboon skins. He carried two long knives and on occasion cut meat from his victims and ate it. Another lion-man was an imbecile who was believed to have had his tongue removed.

The court was told how children were either kidnapped or sold to witch doctors, who kept them in dark, underground grain stores where they were unable to stand upright. They gradually developed a crouching walk and were never again capable of walking erect. All of them were eventually driven insane. Their wrists were broken and their hands were tied back against their forearms to simulate pads of animals. The tendons in their legs were cut to give them a particular gait. The court heard how a fifteen-year-old girl watuSimba—kidnapped at the age of eight—was hired out as a killer at four dollars a time. Some of the lion-people lived on a diet of meat and lived in a lair like animals; they even copulated like animals (two were women). Almost every night they slunk out of their dens in search of random victims.

Some victims may have been almost entirely consumed, for a five-year-old girl, snatched from her mother by a Singida lion-man, was found. Only her skull, teeth, and a few other parts were left. Three witch doctors—two of them women—were hanged for this particular killing, but the lion-man himself was never discovered.

If lions in the wilds are, under normal circumstances, predictable, intelligent, and shy, this is certainly not the case when it comes to lions in captivity. Captivity can often dull intelligence in an animal and in lions it appears to have some fairly profound effects. No longer do they need a keen sense of smell or hearing, nor do they have the opportunity to compete for territory (an animal's most basic preoccupation, if we are to accept current theory). No longer do they have to hunt. It is little wonder that Colonel Stevenson Hamilton found in Britain that a lion born into a long line of captive lions, either in a zoo or in a circus, often has a brain case 50 percent smaller than that of a wild lion.

It seems that circus lions, whether they are trained in the old way or with animal psychology, are either passive, couldn't-give-a-damn lions or mean and dangerous beasts. About two dozen circus men have died in lion acts in this country and a number of zoo attendants have died. Hans Brick, whose animal-taming father was killed by a circus tiger, also preferred to capture his recruits in the wild, for he found wild lions held man in respect as a superior animal. Dick Chipperfield, too, frequently went to Africa to capture young lions for his circus. Today his family—now living in South Africa—owns five hundred lions. Chipperfield claims that "99 per cent of all tamers killed by lions in the ring were victims of male lions." He told me: "I very seldom have a male in the ring with a female. The male is liable to take you for a rival and he will hardly miss an opportunity to get you. Male lions easily misunderstand one's motives—they are nowhere near as intelligent as lionesses. Lionesses accept you as a male lion and I have heard of lionesses —indeed I have seen them—take the side of the trainer who is threatened by a male lion."

A common cause of death in the ring is a trainer's backing into an animal or treading on its tail or foot. On March 31, 1932, Chipperfield saw his own brother-in-law, "Captain Purchase," misjudge an action in a small cage in the ring and confuse a male lion. The lion sprang at him and began to tear at him as he lay helpless in the sawdust. Chipperfield, then aged thirty-six, slipped into the cage as the audience began to scream hysterically, and attacked the lion with a feeding fork. He got the fork into the lion's mouth and began to jab it into its throat. Purchase was dragged out by attendants but died three weeks later from infection. Infection was responsible for several deaths in circuses in the early days and quite often a clawing from a lion could carry such potent bacteria that it was tantamount to being bitten by a venomous snake.

Chipperfield himself—now over seventy—might have died many times had it not been for antibiotics. "I have lost a lot of little bits of bark," he said.

The ex-rector of Norfolk, Reverend Harold Davidson, who

gave up the cloth to become a showman, died in a lion act at the age of sixty-two. One day, in Pleasureland, Skegness, England, Davidson stood on a lioness's tail and as his horrified audience looked on the lioness clawed him to the ground and then carried him in its mouth to a corner, where it killed him.

A number of captive lions which go berserk attempt to eat their victims. Hans Brick once had a most cooperative lion which he himself captured in West Africa. Within a short time of its capture the animal consumed a groom, who, coming home drunk, had staggered against the bars. "He was killed and devoured through the bars," says Brick. Only a few pieces of cloth and a shoe containing a foot were recovered.

In 1966 a show lion—almost definitely born in the wilds —escaped near Kampala, Uganda, and killed seven people before it was shot. In odd places circus and zoo lions have escaped from time to time but rarely with such disastrous results. Usually, once a trained lion escapes it frantically seeks a place to hide. Guggisberg gives an excellent example of this in *Simba*. Somebody telephoned Dr. Gebbing, former director of the Leipzig Zoo, to say that an escaped circus lion was sitting on the roof of a building. Gebbing advised them to get its cage and put it near the lion and then move in on the lion. It will seek the safety of its own cage, the doctor assured them. This is precisely what happened.

Lions have played a bloody though relatively insignificant part in history, unlike the elephant, which, used in war, changed the fortunes of empires. Lions too were used in war, but more for their dramatic appeal than for any real strategic values. Rameses II would invariably go to battle in a chariot flanked by his special lion, Auto-m-nekht, which effectively kept people away from him. The Assyrians and Babylonians used lions too, both for war (mainly as bodyguards) and for eating prisoners, a spectacle usually open to the public.

Lion hunting in biblical days must have been costly in terms of lives: Tiglathpileser I "at the command of my patron . . . killed 120 brave-hearted lions in heroic battles on foot, 800 lions from my chariot." Assurnasirpal II boasted of having killed "370 lions like caged birds" with a spear. The Greeks in Homer's period were fond of reckless lion hunts and according to Homer Greece was plagued with man-eating lions which had probably been captured in Asia and set free in Greece for future hunts.

Although the use of lions in war was neither popular nor particularly successful, the same cannot be said of lions used for execution purposes and in Roman arena events. In 185 B.C. Marcus Fulvius Nobilior ordered that all lions and leopards captured by Romans were to be killed in arena events. Julius Caesar celebrated the consecration of the Forum by having four hundred lions slaughtered by various means, although mainly by men. Nero had his mounted guard, armed with lances, fight three hundred lions in the arena. It was Nero who first had Christians executed by lions. Half of Rome had been destroyed by fire and Nero blamed the Christians. In an orgy of blood-lust he had Christian prisoners dressed in animal skins and put into the arena, where they were killed and often eaten by lions.

Gladiators too were killed in large numbers by lions, which were invariably starved into a state of frenzy before the contests. Some gladiators wore a gauntlet, carried a shield, and used a spear, or occasionally a sword or trident. Some faced the lions wearing nothing and armed with only a net or cloak, with which they were supposed to confuse and then strangle the lion. These desperate men were recruited from the ranks of criminals, war prisoners, or young men in need of money. They were well trained and apparently took pride in their status. Some fought lions with such courage that, if they were criminals or prisoners, they were freed. From the time they were introduced into the arena (around 264 B.C.) until man-versus-animal events were abolished some five hundred years later, thousands upon thousands of gladiators and helpless prisoners were killed and occasionally eaten by lions before fascinated audiences. And if the humans could be counted in thousands, the lions that were killed could be counted in tens of thousands. It was solely the enormous toll of lions killed in the arena that caused them to become extinct in Mesopotamia and throughout many other areas exploited by Rome.

Strangely, the Romans felt a little sentimental about the lions that ate their prisoners. They would get upset and restive if a lion was hurt or made to look ridiculous by men fighting for their lives in the arena. Pliny, who, like most of Rome's intelligentsia, was never happy with Roman "blood sports," wrote of the lion: "The lion alone of all wild beasts is gentle to those who humble themselves to him and will not touch any such upon their submission, but spares whatever creature lies prostrate before him. Fierce and furious as he is at other times, he discharges his rage upon man before he sets upon woman, and never preys upon babies unless it is from extreme hunger." His theory is magnificent humbug.

Conrad Gesner, who wrote an animal book in German in 1563, is quoted by Guggisberg: "So peaceful and mild is the lion, that he does not wound anybody who throws himself on his mercy. Nor does he harm anybody more than he has been harmed himself . . . does not hurt man at all if not driven to it by great hunger, and even then he touches women less than men and spares children altogether, such a righteous animal is he." Magnificent humbug, did I say? In 1960 a European woman in Kenya was walking along a track near her home when she saw, coming toward her, a lioness with cubs. The mother became rooted to the spot. Now here can be a tricky situation, for lionesses are extremely nervous and usually aggressive when with cubs. The lioness advanced upon the frightened woman, who was carrying one child and leading another by the hand. The lioness pulled up just short of the trio, eyed them in the dispassionate way that some cats adopt, and then returned to her cubs and withdrew them from the track. The incredulous mother walked by. What made the lioness so understanding? Perhaps it is true that if you stand still when charged by a lion it will lose courage. Perhaps.

The Tiger

"When the stars threw down their spears,
And water'd heaven with their tears,
Did he smile his work to see?
Did he who made the Lamb make thee?

"Tiger! Tiger! burning bright
In the forests of the night,
What immortal hand or eye
Dare frame thy fearful symmetry?"

William Blake

The most people recorded to have been killed by a single feline were the 436 killed and eaten by the tigress of Champawat. The tigress was hounded out of Nepal in 1907 after accounting for the deaths of two hundred people in four years. It then entered the hill country of Naini Tal in northern India, right up against the frontier dividing the two countries.

The Champawat tigress immediately embarked upon a man-eating career that spread over the next four years and cost a further 236 lives. Her toll was so enormous for such a relatively short period, and her kills so regular, that it appears she lived almost entirely upon a diet of human flesh.

Living in Naini Tal at the time was Jim Corbett, who had been born there thirty-two years earlier and who, in the next thirty-two years, was to eliminate a dozen man-eaters—tigers and leopards which, among them, had eaten 1500 humans. The hunter left India when the Old Order was demolished by Partition in 1947. From then he lived in Nyeri, Kenya, where he died after a heart attack in April 1955 at the age of eighty.

Corbett's first contact with the tigress came after he was called to Pali, a village between Dabidhura and Dhunaghat, where a woman had been clawed out of a tree by the tigress, which had crept up on her while she was cutting leaves. In her desperate effort to hold on to the tree as the tiger took her by the ankles, she left the skin of her palms on a branch.

Corbett arrived at Pali five days after the killing and found its fifty inhabitants still living behind locked doors. They told him that for three nights after the woman had been taken they heard the tigress roaring one hundred yards away.

As the majority of people in these parts were Hindus, a great deal of importance was attached to finding the remains of a tiger's victim—even if it were only a sliver of bone—so that the remains could be cremated, thus ensuring that person's entry into heaven. One wonders how many relatives died because they went searching for the remains of a loved one. The man-eater, on this occasion, had left a few splinters of bone and some clothing.

While in the area Corbett met a woman who had actually chased the tigress a year before after it had run off with her sister. Armed with a sickle, the surviving sister raced after the tigress which, dropping its dead victim, charged her. The sister then turned and fled and upon arriving home she was found to be dumb with shock. Corbett claims that she did not recover her power of speech until he was able to show her the carcass of the tigress.

Fifteen miles east of Pali is the village of Champawat. On Corbett's arrival there he heard a story from a party of men which illustrates the horror and the strength of a man-eating tiger. The men told Corbett that they were on the path that led to Champawat fifty yards above the valley when they "were startled by hearing the agonized cries of a human being coming from the valley below. Huddled together on the edge of the road we cowered in fright as those cries drew nearer and nearer, and presently into view came a tiger, carrying a naked woman. The woman's hair was trailing on the ground on one side of the tiger, and her feet on the other—the tiger was holding her by the small of her back—and she was beating her chest and calling alternately to God and man to help her.

Fifty yards from, and in clear view of us the tiger passed with its burden, and when the cries had died away in the distance we continued on our way."

While Corbett was at Champawat the tigress made its 436th kill. It took a girl, sixteen or seventeen years old, who had been gathering sticks. Corbett raced to the scene and found only a pool of blood, a broken necklace, and the pug marks of the man-eater leading to a ravine choked by jungle growth and rocks. He followed the blood trail and found first a sari and then a bloodstained shirt. Deeper in the ravine he found where the tiger had stopped, only minutes before, to begin its meal. Corbett had disturbed it and it had abandoned one of its victim's legs, which had been bitten off below the knee "as though severed by the stroke of an axe." While looking at the leg Corbett, quite new to the deadly game of seeking out a man-killer, had forgotten all about the tigress. He suddenly felt in very great danger. "Hurriedly grounding the butt of the rifle I put two fingers on the triggers, raising my head as I did so, and saw a little earth, from the fifteen-foot bank in front of me, come rolling down the steep side and plop into the pool." The tigress had been watching him.

The intrepid Corbett passed on. By now the tigress was beginning to resent its pursuer and she snarled each time he caught up to the spot where she had hoped to settle down to eat. Corbett admits to being terrified and after a time he wisely abandoned his quarry—temporarily.

Corbett left the man-eater in a rugged tangled valley through which ran a stream. Looking down into the valley from the hills that form its southern side, he saw a narrow gorge on the right. The left-hand side of the valley was fairly open. Here was a perfect situation for a beat and Corbett wasted no time in organizing one. He rounded up 298 volunteers who, in spite of their fear of the tigress, agreed to block the open end of the valley. They were armed with old muskets, sticks, boxes, stones—anything that could make a noise. Corbett briefed the beaters to keep quiet until they saw him signal with his handkerchief. The hunter felt duty bound to take the local headman with him and the two men began to walk along the face of the hills just above the valley floor. The headman said his shoes were hurting him and sat down to remove them; the beaters took this as a signal to begin. In a second the valley echoed with the dreadful din of 298 frightened and excited men firing guns, beating sticks, hurling stones, and shouting and screaming. Corbett raced headlong down the slope and took up position just in time to see the man-eater emerge from a thicket three hundred yards away. The headman, quite unexpectedly, fired both barrels and the tigress whipped around and went straight back the way she had come. Corbett threw a despairing shot after her but he must have realized its hopelessness. It seemed the hunt would end with the tigress ripping her way through the line of beaters and Corbett sat and waited for their renewed screams. Instead he heard them roar with triumph and fire off more shots: the beaters, having heard the three shots, had assumed the tigress of Champawat was dead. The din had a fortunate effect: the tigress retreated before it and headed straight for the narrow gorge where Corbett waited. As she came into view he took a bead with his .500 modified cordite rifle, and fired. The great cat stopped, then slowly she turned her head in his direction. The moment of truth, as hunters say, had arrived. Corbett fired again, a shoulder shot. The tigress began to thresh at the bush—but Corbett could do nothing now; he had no more cartridges. He usually carried only three—"one for spare." He called the headman to give him his gun but the headman called back that his feet were sore. Corbett rushed up the slope to where the headman was sitting, snatched the rifle from him, and raced back.

"As I approached the stream the tigress left the bush and came out on the projecting rock towards me. When I was within twenty feet of her I raised the gun and found to my horror that there was a gap of about three-eighths of an inch between the barrels and the breech-block. The gun had not burst when both barrels had been fired and would probably not burst now, but there was a danger of being blinded by the blow back. However the risk had to be taken, and, aligning the great blob of a bead that did duty as a sight on the tiger's mouth, I fired." He missed. The bullet, with a velocity that matched that of a potato gun, struck the tigress's right paw. It hardly penetrated, for Corbett was later able to remove it with his fingers. But it was enough: the man-eater of Champawat fell gently forward until her head rested upon a rock. She was dead.

During the course of this century tigers have hunted almost the length and breadth of the Asian continent from the Caspian to the Sea of Japan, from Southeast Asia to the frozen wastes of Siberia. Over tens of thousands of years they were as much at home padding through the snows as they were in the desert; they stalked the steaming jungles with the same success that they stalked the brittle plains. Few other animals have shown such versatility of environment.

Today the tiger is retreating. Apart from a few score surviving in Iran and maybe a hundred strictly protected in the Soviet Union, they have been shot out over five million square miles of western, central, and eastern Asia. Only seventeen thousand are left in the world today. According to Richard Perry there are "a few hundred" left in China; "less than 4,000 scattered over India . . . and perhaps a further 12,000, perhaps more, throughout Nepal, Bhutan, Indo-China, Burma, Thailand, Malaya and Sumatra."

Like the rest of the world's spectacular animals the tiger's population has been decimated beyond the point where it can hope to recover anywhere near totally. Perry is pessimistic about its chances of surviving even to the end of the century.

There are many people who advocate the tiger's extermination on the grounds that it is still a very formidable killer of people. Armand Denis, who points out that a single tiger has caused the destruction of thirteen villages and the desertion of more than 250 square miles of once productive land, says: "One can see the practical reasons for a policy of extermination." Extermination, however, is an irrevocable step which has seldom if ever proved justifiable. There are very good grounds for believing that the tiger has in fact saved more human lives

than it has taken—in spite of stories like the Champawat
tigress. Throughout India one frequently comes across communities with a strong respect for the tiger, a respect born not
so much of fear but of a strange gratitude. Peasants recognize
the tiger as a crop protector, for its main diet comprises such
crop raiders as wild pigs, goats, and deer. Denis estimates that
in addition to the jungle creatures it kills it also kills thirty
thousand head of cattle a year, but even this is mitigated by
the good it does. The belief that the tiger saves more people
from starvation than it consumes for its own welfare has often
led to peasants' tolerating a man-eating tiger in their midst.

The peasant attitude toward tigers is indeed a strange one.
In Burma they refer to the animal as *saya,* which means "master."
In some parts of the tiger's range, especially in parts of India,
people live in abject fear of it while in other parts, even where
the tiger is common, the big cat is less feared than the leopard
and the elephant. But if no tiger terrorized an area quite to the
extent that one or two leopards did then there were certainly
many that came close.

Most hunters, at some stage, seem really to get to know their
quarry and their views are seldom at a variance. With tigers it
is different; it is most noticeable how the views of the great
tiger hunters vary. There are some who look upon it as a stupid
animal. They point out that unlike the man-eating lion, which,
once it turns man-eater, changes its habits, the tiger does not.
It still hunts by day. It continues to stick to the same old tracks.
It continues to roar when it feels like roaring. On the other
hand, there are men who see the tiger as a wise and noble
animal. Corbett described it as "a large hearted gentleman with
boundless courage," but then Corbett was a little sentimental
about tigers and in his retrospective moments tended to forget
the appalling misery that they cause from time to time.

Other authorities speak of the tiger's friendliness and even
its sense of humor. Richard Perry mentions how a Malay police
sergeant, sitting at night on a riverbank, was "kissed" by a man-
eating tiger, which sneaked up from behind. "Have you for-

gotten that night on the river bank when the tiger kissed me on the cheek, Tuan? He could have killed me easily had he wished. He did not do so. I am friendly with tigers and do not care to gaze upon them when they are dead. I would prefer not to help anybody kill one."

Where a tiger's actions might allow a number of different interpretations, its appearance allows no equivocation: here is an almost perfectly designed killer. It has power, it has stealth, it has an excellent camouflage and its armory is second to none among the big cats. The tiger is larger and heavier than the lion—which makes it the world's biggest cat—the adult male averaging 9¼ feet from nose to tail tip. The largest ever shot was over 10 feet. It can stand 3 feet at the shoulder and its girth can be anything up to 5 feet. Its legs are much heavier than those of the lion and its weight is usually at least 100 pounds more than a lion of comparable age. Armand Denis records a tiger of 645 pounds, and Nikita Khrushchev was presented with one of 700 pounds a few years ago. A tiger's talons can be anything from two inches to four inches long and its canines measure, on the average, 2½ to 3 inches for the upper jaw and about ½ inch shorter for the lower. One monstrous tiger fang was 5½ inches in length and 3½ inches in girth.

The tiger must be one of the most "humane" man-killers; at least there is some comfort in the thought that many of its victims never knew what hit them. The strength of the animal is enormous and one gets some inkling of it in Perry's story of the tiger which dragged a 1700-pound gaur nearly fifteen yards—a weight that thirteen men failed to move even a yard. H. R. Caldwell once saw one trot half a mile with a 200-pound hog in its mouth. This report is not unique. Tigers are able to clamp their jaws around the hips of a struggling man and carry him off the ground quite easily. Stories exist also of people being decapitated by a single blow from a tiger's paw.

The African lion, in popular fiction at any rate, has a far worse image than the tiger and yet if either animal deserves a reputation for man-eating it is the latter. The entire gang of Tsavo man-eaters could not match the misery, death, and disorder created by the tigress of Champawat. Many other individual tigers did more damage than the Tsavo lions. The man-eater of Thak in India, for instance, caused peasants to flee the village of Thak and kept a forestry labor force of fifteen thousand Indians in a state of terror. It seized victims from camps spread over fifty miles apart.

Schaller estimates that a tiger requires $3\frac{1}{2}$ tons of meat a year in order to keep healthy. This means that if a tiger lived solely upon human meat it would have to kill nearly sixty adults a year. Generally, however, when a tiger turns man-eater it tends to treat man as a delicacy rather than a staple and will continue to take cattle and deer when these present themselves.

The hungriest tigers waste very little: they frequently eat everything, including the blood-soaked clothing. Some will leave the brain pan which is licked clean. Some tend to leave the hands and feet intact.

Richard Perry was not exaggerating when he described the tiger as "the monster that for four centuries certainly, and no doubt for very much longer than that, has darkened with terror the lives of millions of Indian villagers." Perry goes on to say that "in terms of human misery it was the constant suspense of being defenseless against sudden and horrible death by day and night, over a period of months or even years, that broke the spirit of the [Indian] villagers, rather than the actual number of deaths among them, though the latter were terrible enough."

What was "the actual number of deaths"? There have been some carefully calculated guesses from time to time and, less often, some official statistics. There have also been some fascinating exaggerations. A. D. Dunbar Brander states: "At one time in parts of India in the nineteenth century man-eaters were so numerous that it seemed to be a question of whether man or tiger would survive."

It gives one pause, but Brander, who hunted in India when man-eating was probably at its peak, must have seen some miserable sights. At a conservative estimate tigers have consumed well over half a million Indians in the past four centuries. In the whole of Asia the figure for the same period cannot be less than a million. "There may have been 300, perhaps as many as 800, operating at one time," surmises Richard Perry. Terrified people passing through some of the plagued areas had to be preceded by men beating drums and carrying fire-brands. They would shout and howl as they passed along jungle tracks in the hope of frightening off lurking man-eaters. Quite often armed guards would wait until there were a number of travelers gathered together and then, flanking them on either side, escort them through particularly bad areas. Johnson, early in the last century, described how the villages of Ramghur were deserted and the fields unattended as day after day victims were carried off into the surrounding jungles. "Whenever a tiger carries off a man near a public road," he wrote, "a stick with a colored cloth, or a small, white, triangular flag on a bamboo staff ten to twelve foot high, is erected to warn travelers; and every passer-by throws a stone on to a heap, which soon become large heaps in great abundance." These cairns, or *bag-houts* as they are properly called, were common throughout India in the eighteenth and nineteenth centuries and were placed every quarter of a mile apart along some of the jungle roads of Mysore. Each stone tossed on to the heap represented a nervous prayer that the traveler would make it safely to his destination. A lot were thrown in vain.

Not until after the Second World War was there any significant drop in the toll, and then it must be remembered India's tiger population had by then been reduced to a few thousand. (It is estimated that a hundred thousand tigers have been shot out in India this century.) The number of mortalities caused by man-eating tigers in the late 1940s dropped to about eight hundred a year.

There are still some impressive man-eaters in India. In 1966 they made headlines in the outside world four times. In January a man-eater at Ramgiri Udaygiri, said to have eaten 500 people in six years (this figure, which appeared in newspapers at the time, turned out to be a gross exaggeration), was shot by Mrs.

Alida Sverdsten of Idaho; in May the police were called in to help hunt down "man-eating tigers" which ate 75 people in two areas of northern India (many of the victims were women, says the report); in June tigers in the Ganges Delta were said to have killed 46 people "so far this year" and in November a tiger in the Kalser district, one hundred miles out of New Delhi, was reported to have eaten 6 children.

Along the mouth of the Ganges, man-eating tigers have preyed over the centuries almost without letup. The 46 people reported killed in the area in the first half of 1966 represent a figure slightly higher than the average experienced in that green and tangled area. Perry estimates that 50 a year die in these parts and says that jungle workers are given firecrackers to explode to frighten off the man-eaters. There is evidence that the firecrackers have the opposite effect and attract man-eaters.

It is clear that no other part of Asia was so persistently and thoroughly picked over by man-eating tigers. Singapore had a rough time toward the end of the last century when 600 to 800 Chinese coolies were taken annually (even in 1929 there were 15 casualties attributed to tigers), but elsewhere the tiger has lived and still does live more or less amicably with man.

Caldwell maintains that man-eating in China was never prevalent, although there were isolated periods when tigers became a nuisance. Right up to the end of the last century China and eastern Russia had a large tiger population but the animals were, for some reason, small in stature and preyed mainly upon children. Thus in 1922, when there was a sudden and short-lived spate of man-eating on the Chinese coast of Fukien, the sixty victims were almost all juveniles. One of the worst areas for man-eating tigers was the taiga and in 1891 Manchurian peasants ventured out of doors only when it was absolutely necessary because of the presence of man-eaters.

Occasionally Cossacks were called out in parts of the taiga to watch for man-eaters when labor gangs were building roads or railway lines. What may have given rise to the greater number of cases in the taiga was the use of "tiger trees"—trees to which thieves were tied for "The Great Van" to find and punish.

The Great Van normally had stripes and having fulfilled his function as a dispenser of justice passed on with new and unhealthy ideas.

Although man-eating outside India was never really a constant problem, the occasional visitations of wayward tigers in the Far East did give rise to a hunters' guild, a group of startlingly reckless men who were called in whenever a troublesome tiger appeared. The nineteenth-century White Russian hunter Yankowsky came across a Korean member of the guild who was using an old-fashioned matchlock. The Korean's companion was crouching next to a still warm tiger drinking its blood.

The tiger was an old male, records Yankowsky as quoted by Perry, and it was of colossal proportions, having eaten several men. The Korean hunter described to the Russian how he had killed the tiger and his blasé story stands out as one of the most hair-raising accounts of the killing of a tiger in literature. "I was here by the kill, behind this tree. I thought I saw two brown birds moving about in that bush, then I realized they were not birds but a tiger's twitching ears. Ho-rangi [tiger] had seen me. At once I blew hard on my fuse cord and put it in the fork of the trigger, and we stalked each other round this tree, I hoping all the time that the gun would be ready to go off before the tiger attacked. When I pressed the trigger the first time, the gun did not fire. The tiger was so close that I feared he would spring. So I shouted in his face, and waved my gun like a stick. Tigers do not like the human voice. He drew back, opened his mouth to roar, and I pressed the trigger again—the gun obeyed, and the tiger rose up so high in the air that he fell over backwards."

With the possible exception of man-eating tigers, tigers dislike the frontal attack when approaching a man and most prefer to ambush. In fact, although a charging tiger is a terrifying sight, it is apparently much easier to turn than a lion or a leopard. A single shot fired into the air is often enough to pull it up in its tracks as it comes bounding forward in that strange, slightly slow motion charge. Even a shout can sometimes make a tiger skid to a halt in the middle of a charge. If a tiger carries through its charge it usually presents the hunter with a fair chance of felling it. Several hunter-authors remark on how massive the tiger's head appears as it approaches but most resist any temptation there might be for a brain shot. The favored shot is in the chest where it joins the neck as this smashes the arteries leading through the neck and may also go through the heart and lungs. A few hunters wait until the tiger has completed its charge and has momentarily stopped to gather itself for the spring.

A. D. Dunbar Brander offers the advice that if one is unarmed or otherwise unprepared when charged by a tiger it is best to stand one's ground. Very often the tiger—provided it is not a man-eater—will lose its nerve and pull out of the charge. Should a man turn and run he will almost certainly be pulled down, and very often a charging tiger has charged one man and then suddenly swung off to attack another whose nerve cracked, causing him to try to run for it. This is a common characteristic among dangerous animals, especially the lion, elephant, and rhinoceros. There are two main reasons for it: the animal's attention is diverted by sudden movement, and it has no need to fear a creature that is fleeing.

The tiger is not a courageous creature. Anderson described it as "craven," and there are instances that makes one wonder if the tiger has not got nerves like violin strings. Perry tells of a tiger which treed three Indians when one of them, in his fright, dropped a blanket on it. With a startled "Woof!" the tiger bolted. Circus men also find the tiger a nervous animal, it might be significant that it was not used much in Roman arena events.

In attack the tiger's forte is the ambush. Quite often it is able to kill its victim before he has a chance to shout or even to know what exactly has got him. Man-eaters use this technique more often than not, and Burton describes it graphically: "An officer walking along a jungle path with his shikari behind him, heard no sound except a deep sigh, and in turning round saw a tiger with a man's neck in its jaws, standing on its hind legs, its forepaws on his neck and chest or shoulders. In an instant it had gone with its prey."

Jim Corbett claimed that once a man is in the grips of a tiger his chances of surviving are one in a hundred. But the same hunter does mention a case in which a man successfully used a stomach throw on a tiger.

W. Robert Foran puts the tiger (a normal tiger, not a man-eater) on a par with the African elephant as the most dangerous animal to hunt. This statement is difficult to justify, but the opinion of such a hunter as Foran cannot be lightly dismissed. Foran may have been unlucky with tigers and, if his first encounter with a tiger is anything to go by, then his attitude is understandable. In this encounter—a hunt from the backs of elephants—Foran was thrown from the howdah and was on the ground when the tiger leaped on to a friend's elephant and tried to tear the mahout from his neck. The mahout, badly mauled, belabored the tiger with his ankus, but this failed to impress the enraged tiger, which clung to the elephant's head. Foran risked a chancey shot and the tiger slid to the ground, where it was trampled and tusked by the elephant.

Some of Foran's beliefs about the savagery of the tiger, however, are contradicted by other observations he makes. One of them concerns an army officer who shot three hundred tigers with little trouble and ended his days by dying peacefully in bed. Many of Foran's stories underline the fact that the tiger, under normal circumstances, is as amiable as the lion and, like the lion, will avoid man twice as carefully as man will avoid him. There are times, though, when the tiger is overcome by curiosity and will approach a man to within a few feet. There are a few accounts of astonished and alarmed mothers who have watched tigers approach from the jungle fringe and lope over to their playing children. The tigers, having stood and watched them for a time, pad softly back to the jungle. On occasions tigers have trotted right up to people and sniffed them as if unsure what sort of animal they were.

In a set of children's general-knowledge books is a drawing of a tiger in a tree and the reader is asked what is wrong with the picture. The answer, according to the editors, is that tigers cannot climb trees. Neither the publishers nor the readers can be blamed for believing this to be correct, for many hunters believe it too. In fact, one of the most popular ways of hunting a marauding tiger is to build a machan "in the safety of a tree" on a path the tiger is known to frequent. The fact is that tigers can climb trees and do climb trees. Perry claims they do it as nimbly as a leopard and have been seen sixty feet above the ground. K. Singh once released a captured man-eater into an enclosure and, after the tiger had charged a dummy, it rushed up a tree, the first thirty feet of which were branchless. In mentioning the incident Perry remarks that many an old machan shot must have broken into a cold sweat when he saw how fast the tiger climbed. There are a few instances of tigers taking men from trees, and Kenneth Anderson was once bitten severely in the seat of his pants by a tree-climbing tiger as he sat in a machan. A man-eater which Anderson and his son Donald shot one moonlit night had also chased a victim into a tree and dragged him down.

The tiger's leap is another of its physical accomplishments that some hunters tend to underrate. These heavily built cats have been known to leap onto the head of a charging elephant and snatch the mahout off its neck or even sweep the howdah off. They can clear eight-foot fences with ease.

Most surprising of all perhaps is its ability to swim well. Tigers have been known to swim up to five miles and on occasion have swum a little distance out to sea. G. P. Evans in *Big Game Shooting in Upper Burma* (1912) mentions how a tiger raced along a riverbank for two or three miles in pursuit of three men in a canoe. At times it plunged into the water and tried to swim out to them. Perry mentions a tiger climbing onto the rudder of a ship anchored half a mile out to sea.

Because of the tiger's extraordinary physical abilities and because of its habitat it can be a most exasperating animal to stalk. Once it enters dense jungle it becomes almost impossible to follow with any reasonable degree of safety. Many is the time that men have walked blindly into the animal they have been spooring or have parted the bushes to peer through and have found themselves nose to nose with a wide-eyed tiger. Fortunately, under such circumstances, the tiger seems to get the bigger fright—or, more accurately, it is the tiger that then jumps the farthest and the quickest.

The most favored method of bringing a tiger to book is to bait it by tethering a buffalo to a stake and then either waiting in a machan overlooking the spot or revisiting the buffalo from time to time in the hope of catching the tiger while it is eating.

But baiting a man-eater can be difficult because quite often a man-eating tiger will go through a period when it will not look at cattle. Exceptional man-eating tigers may concentrate only on a diet of humans. Hunters are sometimes able to sit up over the remains of a man-eater's meal but such circumstances are rare for two reasons. A few man-eaters never return to an unfinished feast because they have learned to be cautious, and a great many man-eaters leave nothing or almost nothing behind. A man-eater that does not fear pursuit will almost certainly eat the entire corpse. Kenneth Anderson found that most of the man-eaters he came across had the habit of leaving the head, arms, and legs intact, or at least the head, hands, and feet. He and his son, when they shot the man-eater of the Crescent Mountains, sat up over the gruesomely deformed remains of a leper—it proved to be the last victim of the Crescent Mountains killer. The remains comprised the feet, hands, and buttocks, and a small part of the chest. The tiger had severed the head some distance back and had discarded it.

Because of the rugged nature of most of the areas where the tigers roam it is often possible, by employing beaters, to trap a man-eating tiger in a valley and drive it toward waiting marksmen, as in the hunt for the man-eater of Champawat. Occasionally it is possible to ambush a tiger by gauging when it is likely to use a certain track and then lying in wait. This is not quite as haphazard as it sounds, for man-eaters will normally make a few kills in one district and then, when the populace has learned to avoid it, move to another district. It might make the rounds of several districts miles apart but it will always return—and almost always it will use the same track. There are so many other methods too: Jim Corbett, working on the assumption that a particular man-eater looked out for women, dressed himself in a sari and successfully lured a man-eater to its death.

It was Corbett who said that man-eating tigers preferred to eat women, at least in northern India, because it is mostly women who go into the forests alone to gather firewood or down to the rivers for water. He said that woodcutters were the second favorite on the tiger's menu because man-eaters were attracted by the sound of chopping, and woodcutters usually work alone, which makes things easier for the tiger. Anderson, working on the theory that the sound of chopping attracts man-eaters, sat in a tree every day for seventeen days, tapping it with his foot to simulate the sound of an axe. In this way he lured the man-eater of Yemmay Doddi to its death.

Some authorities maintain that all tigers would be man-eaters if they did not fear human beings so much. It is difficult to reconcile this theory, for there is ample evidence that tigers living unmolested in their jungles are more curious about a man than afraid of him. A normal tiger that has had to kill a man in order to get at his cattle will rarely feed off the dead man. The Indian tribe appropriately called the Dares used to rush at tigers as they fed at a carcass so that they could obtain meat for themselves. The results were sometimes fatal for the Dares, but there is no indication that the tigers then ate their human victims.

From an ecological point of view it is difficult to see how the tiger could have ever counted man as a serious food source until fairly recently. There would not be much point in tigers eating men when the jungles were filled with buffaloes, deer, and pigs, which the tiger was well equipped to catch. Predators tend to go for the most prolific species, and man was never that until quite recently.

It is possible—indeed Perry says it is probable—that the peak of man-eating in India was in the last century, when it seems that the tiger population reached the saturation point and man's population was exploding. One must also bear in mind that man was moving more into the jungles and that the firearm was becoming commonplace. The last century's chikars probably created more man-eaters than they killed by wounding tigers, and the sudden surge of game-hunting in India caused the buffalo and deer populations to shrink while the peasants' cattle population was expanding. It left the tigers little choice but to learn the ways of men so that they could better kill his cattle.

Even so, a lot of peasants did not take the tiger seriously, and small herdboys would often shoo tigers away from the herds. Old men and women would occasionally lose their tempers when they saw a tiger attack a prize head of cattle and would chase after the cat, which would invariably disappear back into the jungle.

Corbett states emphatically that nine out of ten man-eating tigers are sick and the tenth too old to hunt its natural prey. Osmand P. Breland is equally convinced that nine out of ten are *not* sick. Only one in ten is sick, he says, and the rest are usually in good health. E. P. Gee favors Corbett's theory and says that out of every thousand tigers, three or four are man-eaters and these are either old or sick. Both Breland and Corbett are being unnecessarily dogmatic. Perry notes that "more often than not" nineteenth-century man-eaters were described as old and mangy, whereas modern man-eaters are in much better condition; perhaps they inherited their taste for human meat from their parents and the trait could even go back to some distant ancestor. Anderson recalls that Chitaldroog was plagued for generations by man-eaters and there is a strong suggestion that the man-eating habit was being passed down from tigresses to cubs. Armand Denis, too, records that tigresses pass the habit on to their cubs. On the other hand, Corbett denies that

this is so. He says in an author's note in *Man-Eaters of Kumaon:*
". . . the reason why the cubs of a man-eater do not themselves
become man-eaters is that human beings are not the natural
prey of tigers.

"A cub will eat whatever its mother provides, and I have
even known of tiger cubs assisting their mothers to kill human
beings; but I do not know of a single instance of a cub, after
it had left the protection of its parent, or after that parent
had been killed, taking to killing human beings." Corbett's
argument is patently thin.

Many man-eaters throughout India and Malaya took up
man-eating accidentally. A familiar pattern would be for a
tiger to kill a herdboy who got in the way when the tiger was
after, say, a cow. The tiger would be unlikely to touch the
boy. A few days or weeks later it might have to kill a man
again. This time there would be less hesitation, but it would
probably not try to eat the man. It might sniff at the corpse.
Perhaps on the third time it is challenged when it attacks
a herd, it would kill with impunity and it might well take
a bite from its victim or perhaps merely lick up the blood.
Having lost all fear now of this funny, two-legged creature, and
being hungry, the tiger might well finish eating its victim. From
then on it might kill whenever the opportunity arose. Another
man-eater might begin its career mistaking a herdsman or wood-
cutter, bending down working, for a pig or a deer. Its sense
of smell would not help much, for, according to Corbett,
Anderson, and Denis, tigers have no sense of smell. Gee disagrees
and says its sense of smell is "normal."

Fewer instances of man-eating have been reported in southern
India, according to Perry, because the peasants show less pluck
when defending their herds against tigers. They avoid provoking
them. A significant exception was the one-eyed tiger shot by
Anderson which terrorized northern Coinbatore for three months,
eating several people. It began its short but devastating career
after a peasant set an inadequate gin trap for it.

It seems to be true that most confirmed man-eaters have
fairly silky coats; several hunters have remarked upon this. The

same is true of lions and leopards. The reason for this is obscure, but Corbett mentions that there is a general belief that man's flesh is more salty than game meat. It is quite possible that man is saltier than game but whether this puts gloss into cat fur it is difficult to say.

Mental derangement can produce man-eaters and the Mauler of Rajnagara probably falls into this category. It began its career in 1955 by clawing, but not at first eating, herdsmen. Perry says it never used its teeth but in another account the Mauler is alleged to have eaten some of its victims. It mauled thirty-three, ripping them down from the scalp, using only its front claws. Eleven of its thirty-three victims died of blood poisoning or shock before Anderson was able to shoot it.

Rabies can be another cause of man-eaters, although rabies in tigers was never recorded until 1943, when a young tigress ran amuck in the Nowgong district of Assam, attacking eighteen people in thirty-six hours. Eleven of them died. A second case of rabies was confirmed in 1950, when a tiger killed three and mauled eleven, even breaking into homes to get at the occupants.

Other very frequent causes of tigers becoming man-eaters have been wars, plague, famine, and floods. The tiger, being partly a scavenger, will often gorge on human corpses after a catastrophe. Normally, of course, the Hindus burn their dead, but there are times when this is not possible. In 1942, thousands of Indians were evacuated from Burma through the Taungup Pass of Arakan Yomas. Four thousand people died during the evacuation and the tigers cleaned up most of the bodies. Four years later, when a West African army contingent passed through the area, fourteen soldiers were taken by man-eaters.

The Leopard

". . . red in tooth and claw."
Tennyson

The road to the Albert National Park between Lake Kivu and Lake Albert in Central Africa is a winding road that picks its way through damp, lush vegetation, the occasional banana plantation, and clusters of beehive huts. On the right of the road and soaring 14,557 feet above the jungle is Mount Mikeno, a volcanic peak that, in the swirling mists which usually shroud its head, broods like an unhappy ogre. Through glasses one can just make out on its slopes the great belts of dripping bamboo, chief food of the last of the world's gorillas. If the mists lift one can also see the spot where they buried Carl Ethan Akeley.

A sculptor, taxidermist, and naturalist, Akeley was an American by birth and an African at heart. He made five visits to Africa to collect and to observe. On his return to America he organized the mounting of some of the world's finest wildlife-museum exhibits, including one tableau of an entire family of elephants. It has been said that his imaginative handling of museum exhibits ended the era in which taxidermy was something akin to soft-toy stuffing.

Akeley died a natural death at the age of sixty-two in 1926, but twice he had cheated death—once when an elephant badly mangled him and once, in 1896, on his first visit to Africa, when a wounded leopard charged him at point-blank range and sent his rifle flying. He was caught completely by surprise when the cat came at him and had time only to throw his arm up in front of his face. The leopard seized his arm in its teeth and began clawing at him with its front paws. Akeley, with his free hand, grabbed the cat around the throat and held it away from his body, his main concern being that the leopard might bring up its back legs and rip him down the middle. Leopards occasionally use this tactic, although in most attacks they prefer to latch on with their front claws and then tear with their teeth. The taxidermist tightened his grip on the cat's throat and slowly worked his arm out of its mouth every time he felt its jaws relax. Soon only his fist was in its mouth and this he kept there so that the leopard could not bite his face and head. Akeley then deliberately fell on top of the animal and dug his knees into its rib cage and his elbows into the leopard's armpits to force its flailing front feet apart. Slowly the cat went limp. By the time Akeley rose the leopard was dead. It had died from strangulation, and its ribs were cracked from the pressure Akeley had exerted. Akeley's escape is by no means unique—wounded leopards, ferocious though they are, rarely manage to kill their human victims and usually end up running away.

The leopard, whether it is hunting along the nullahs of India or on the African veld, is a veritable Jekyll and Hyde. Unmolested and in normal health it is a shy, almost nervous animal with a very marked fear of man. Unlike the lion and tiger, both of which move out of man's way with aplomb and dignity, the leopard will usually run, perhaps pausing to cast a quick glance over its shoulder. Should you make a sudden movement the cat will spring into the nearest thicket and then flee like a startled hare. Many are the stories that speak of the leopard's apparent lack of spirit.

But when a leopard is wounded, trapped, or cornered, it is an entirely different animal. Irrational, irascible, and utterly ruthless, a wounded leopard will almost always attack the first man to come within striking distance. There is no other animal in the world more likely to attack when wounded, and many hunters believe that there is no animal more desperate or savage. Stevenson Hamilton described a wounded leopard at bay as "the very incarnation of ferocity, his ears are laid back low against a flat-looking head, his long white teeth gleam between withdrawn and snarling lips, while his eyes, fixed with steady and sinister snare upon his enemy, are filled with dull, greenish red light, glare murderous hate. Even when you know his back to be broken, his appearance is so little assuring that you have qualms about approaching close for administration of the coup de grâce."

But before we look at the leopard as a man-killer or as a man-eater let us look at him under happier circumstances, unmolested in his natural habitat. *Panthera pardus* is the most graceful and beautiful of all the big cats and is found throughout Africa from the Cape of Good Hope to the shores of the Mediterranean, including the jungles of Central Africa, where the lion is absent. It is found thinly distributed in Asia Minor and then is more densely distributed in western Asia and in the Far East to the Chinese seaboard and down to Java. Gee estimates its population in India at six to seven thousand. Leopards are also found in Ceylon where the tiger is unknown. Technically the leopard is extant in Europe, for a few are said to survive in Kuban, southeast Russia (Denis) and if it is so these would be descendants of the leopards which hunted Europe in Pleistocene times.

Throughout this vast area the species remains the same with only subspecific differences, which are mainly in size and coloring but which to the layman would hardly be noticeable. The panther of India is precisely the same as the leopard of Africa, and the Asian "black panther" is merely a melanistic leopard which is capable of throwing off normal, spotted offspring. The "tiger" of South Africa is also the common leopard.

The animal is inclined to be a little on the stocky side when conditions are really favorable but never so much that it loses its graceful appearance. It weighs an average of 100 pounds or slightly more and a really heavy one in the wild might go as high as 160. In captivity I have known one of 200 pounds. The record length of a leopard is nine feet, which is not far off the record length for a lion—but a third of a leopard's length is tail.

It is strange that a creature so well equipped in tooth and claw, so given to fierce attacks when wounded, and so capable of lightning action should, at the same time, find it difficult to overpower a man. (I am referring to the "incidental" attackers such as freshly wounded ones and not to confirmed man-eaters, which are a different kettle of fish entirely). The leopard certainly doesn't lack strength. T. Murray Smith saw a 150-pound leopard kill a 90-pound antelope, and carry it in its jaws "like a dog carrying a hare" and then climb thirty feet into a tree with it. This feat is by no means unusual, for leopards frequently carry their prey into trees out of the reach of hyenas, jackals, and lions, and there is a record of one carrying a 200-pound baby giraffe twelve feet up a tree.

When then does the leopard not employ this strength when tackling a man? I can only believe it is because it relies upon fast claw work and biting to overcome its victim—unlike the lion and the tiger, which often kills by cuffing its victim. And it might be partly due to the leopard's streak of cowardice. Perhaps cowardice is not the right word: the leopard's fear of man could be a mark of its intelligence and, along with its strictly nocturnal habits, is one of the chief reasons it has survived within sight of large cities.

This apparent feeling on the part of the leopard that discretion is the better part of valor is often manifest when the cat is attacking a baboon troop and runs into a big male. I once watched such an incident. The male baboon immediately attacked the stalking leopard, and screaming at the capacity of its lungs it tore at the cat's neck and shoulders with its long canines. The leopard, spitting and snarling, rolled in the dust with the baboon and then withdrew and tried to "box" the animal with open claws. But the baboon knew better and once again closed in and began biting. After perhaps a minute the big cat disengaged itself and fled ignominiously, leaving a badly torn baboon licking his wounds.

In the old days a good 75 percent of people mauled by leopards died from infection caused by the rotting meat between the leopards' toes. Gee states that the incidence of infection from leopard wounds was higher than with tigers. With the advent of penicillin, deaths were cut down to less than 10 percent of people mauled.

Going by the very inadequate statistics available from local authorities and wildlife organizations, and mainly from incidents reported in newspapers, leopards probably cause a minimum of four hundred casualties a year. Of these only a quarter would be due to man-eating leopards and the rest to the work of wounded or cornered leopards. Of the remaining three hundred victims of "accidental" skirmishes, only 10 to 15 percent die of their wounds and in most cases the leopard is either shot or killed by hand. Either those that escape die of the wounds which originally made them ill tempered or their wounds mend; either way there are very few leopards that get the chance to turn into habitual maulers. On the basis of these figures, there must be between eight and ten thousand people alive in the world today who bear the scars of a fight with a leopard. The pattern of attacks and their frequency seems no different in India from what it is in Africa.

In the four years ending in December 1964, two Johannesburg daily newspapers, the *Star* and the *Rand Daily Mail,* carried reports concerning attacks by thirty-two leopards in South Africa, South West Africa, Rhodesia, and neighboring Mozambique. This does not mean that there were attacks by only thirty-two leopards in those four years in southern Africa —there were undoubtedly more, especially in remote areas, which went unreported. There were, for instance, six supposedly man-eating leopards shot in the north of Mozambique, but so wild and so lacking in communications is this notorious area that it was months before the incidents became known and by then, one assumes, the news was too stale to print.

None of the reported attacks (some were on more than one person) was fatal; but thirty-one people were mauled, most of them severely, and eight survived unscratched. None of the leopards made a subsequent attack. All of the thirty-two leopards had been provoked by being wounded, or by being attacked by the victims' dogs. Ten escaped and were apparently not heard of again, ten were shot, seven were speared, clubbed, or axed to death, and five were killed barehanded. In four of the attacks the victims were able to punch the leopard on the nose, and in all four of these instances the leopard broke off the engagement (in one of them only temporarily).

In 1960, according to these reports, three leopards were killed barehanded, one of them by a seventy-year-old, gray-haired African named Kudziburira, who took his dog and an axe to look for a leopard which had killed one of his goats in Sinoia, Rhodesia. The dog flushed the leopard and was immediately killed and the cat then turned on the old man, who in his fright dropped his axe. The leopard went for him and bit into his left arm, which he had thrown up to protect his face. The man gripped the cat by the throat with his free hand and held it there until he felt its jaws relaxing. Then, pulling his arm from its mouth, he put both hands around the cat's neck and half-strangled it. He let the leopard fall into an unconscious heap before he picked up his axe and delivered the death blow. Two months after this, Castelo Branco Montero of Lourenço Marques, Mozambique, was attacked by a wounded leopard, which he managed to strangle and not long after another Portuguese, Anselmo Gomes Matos, strangled a leopard with his bare hands at Silva Porto, which is in Portuguese West Africa, just north of South West Africa. In all these incidents the men were severely mauled.

There are scores of incidents on record of men strangling or choking leopards, and in at least one recorded incident the men killed two leopards by crashing their heads together. The man who performed this feat was Cottar, a Texan who became a white hunter in East Africa and who, according to J. A. Hunter, killed three leopards barehanded in his career.

In India leopards also occasionally find the tables turned against them: in July 1963 there was a remarkable story from Calcutta about an elderly woman named Debku who, in the state of Himachal Pradesh, throttled a leopard after an hour-long struggle during which her dogs kept worrying the cat.

Significantly, many hunters class the leopard as the most dangerous of all the big game animals to hunt. Ionides considered it the most dangerous animal in the world, and Rushby classed it as more dangerous when wounded than a wounded buffalo. He described the leopard as "a perfectly built killing machine" and said that as a target it was so small and so fast that it was most difficult to hit. To make things doubly difficult for the hunter, the leopard specializes, a little like the tiger, in short-range charges. But, in spite of all the claims from hunters, I cannot think of a single big-game hunter who was killed by a leopard. Some may well have been, but the number must be insignificant.

There is a theory that a leopardess will charge a man on sight if she has cubs with her. Ted Reilly of Mililwane Game Sanctuary, Swaziland, told me how he once came across a leopard cub in the bush, and assuming it was orphaned or abandoned, picked it up. Even as he straightened up he could sense danger, and as he stood holding the spitting bundle in his arms, he found himself looking into the smoldering eyes of a she-leopard. Reilly, being alone and unarmed, fairly jumped out of his skin. His involuntary movement caused the mother leopard to turn and flee. This behavior seems to be fairly unusual, but it does illustrate how highly strung these animals are in the presence of man. An Acholi game ranger in Queen Elizabeth game reserve in Uganda was not so fortunate the day he cycled past a leopard cub. He guessed the mother would not be far away so he pedaled harder. The next thing he knew the mother was pelting in behind him and sprang onto his back, where she remained firmly latched as the ranger put on an admirable burst of speed. The leopard fell off two hundred yards farther along.

A similar incident befell Arian Suque of Dodoma, Rhodesia, in 1964, when a mother leopard with three cubs attacked him and pushed him off his bike. Suque fought the leopard and she ran off, but a minute later she came back for another round. Again Suque fought her, and again she retreated. But this was an occasion when the leopard came out on top: she punctured the African's wheel and he had to walk ten miles to a hospital.

At the state opening of Kenya's Parliament in 1963, Senator Godfrey Kipbury, the Masai member of the Senate, made a grand entrance wearing a smelly leopard-skin hat. It smelled because the skin was only forty-eight hours old. The senator had speared the leopard on his farm but was clawed in the process. Annually, scores of leopards are speared by Africans mainly because of their threat to livestock.

Although India, through the writings of such men as Jim Corbett, has received more publicity for its man-eating leopards, Africa's record is as bad. Throughout its range the leopard, when it does turn man-eater, goes mainly for children and sick adults, and sometimes it has even dragged humans into its larder in the branches of trees. George Rushby, who was senior game ranger in the notorious Njombe district of the Southern Province of Tanzania, said that at least fifteen Africans a year were eaten by leopards there during the 1950s. Oddly, most of Africa's man-eating leopards have been in Central Africa. I have come across no instance of man-eating leopards at all in South Africa, and there are few in East Africa, but in Central Africa man-eating leopards have caused entire villages to be abandoned. Some leopards are crippled or old, and one feeble old cat which killed twenty-two people in northern Mozambique was even too weak to carry its victims off. It used to tear chunks out of them and then limp away into the night. C. J. P. Ionides describes the depredations of a typical man-eater that haunted Rupenda, Tanzania, in 1950. This leopard might have been forced into eating man because of the ill-fated British

Groundnut Scheme. Africans had been clearing the bush for miles around for the scheme and the leopard's natural prey had entirely disappeared. The leopard began by killing a laborer, but before it could eat him it was beaten off by villagers. That same night, six miles away, it ate a baby after stealing it from its bed. Two days later it hid behind some scrub and watched a mother teaching her toddler to walk, then when the mother went indoors for a few seconds the leopard snatched up the child and later ate it. Trackers found that the leopard must have crouched behind the bush for some time waiting for a chance to spring. A third and fourth child were taken near Rupenda before Ionides got to the village and then the leopard struck a fifth child seven miles away. Eighteen children were taken in a few months; the youngest was a six-month-old baby and the oldest a nine-year-old girl. Eventually the leopard was caught in a trap.

Leopards will often kill humans apparently for the sake of killing. One of the most notorious in this regard was the leopard of Masaguru, Tanzania, on the Ruvuma River. It killed twenty-six women and children, not one of whom was eaten. Rushby says that not even a bite was taken from them; and when, eventually, it was shot, it was found to be in good shape. Possibly the leopard had taken a liking to drinking human blood, for this has happened from time to time.

Rushby has an interesting theory about man-eating leopards and surmised that they eat humans merely for variety. He points out that even as man-eaters they still continue to kill and eat monkeys and baboons. In fact, says the hunter, there cannot be much difference between human meat and monkey meat.

In India there have been some incredibly successful man-eaters. The Panar leopard is supposed to have eaten four hundred before it was shot a generation ago. The man-eater of Rudraprayag ate 125 during roughly the same period. More recently, the man-eaters of Bihar state in 1959 and 1960 are said to have eaten three hundred people. There are so many more: the Gummalapur man-eater, shot by Kenneth Anderson after it had eaten forty-two people; and a host of others that have plagued the wild and beautiful valleys of central and northern India. In centuries gone by, the record hardly varied. Oddly enough, man-eating leopards are almost as rare in southern India as they are in southern Africa—except in the narrow, four-hundred-mile-long forest belt in the Western Ghats.

Individual man-eating leopards in India kill a significantly greater number of victims than their counterparts do in Africa. One reason is that the man-eaters of India seem to survive a great deal longer than the man-eaters of Africa. But this is only half the answer. Is it perhaps because the Indian hunter is less persistent and bold when pursuing man-eaters than the African hunter? Jim Corbett, the man who shot both the Rudraprayag and Panar leopards, puts it down to the way Hindus, in times of plague or famine—when there is no time to burn their dead as they normally do—put a piece of coal into the corpse's mouth and leave it in the bush. Leopards, always partial to a little carrion, eat these bodies and acquire a taste for human flesh. This is not an entirely satisfactory answer, because in Africa tribesmen in many parts *habitually* leave their dead in the bush for scavengers. This does create man-eating leopards, but none has even approached killing the numbers claimed by the big man-eaters of India.

The whole character of a man-eating leopard differs from that of the mauler. Man-eaters show no ill temper but instead hunt with awesome calm and cunning. In India, as in Africa, most of them specialize in children, and E. P. Gee speaks of one which ate only children of between six and ten years. Because of the leopard's strictly nocturnal habits and its custom of hanging about on the outskirts of villages in the hope of knocking down stray livestock or village dogs, the Indian leopard is familiar with the ways of men. To add to its efficiency as a man-killer, it is always supremely cautious, never really losing its fear of man no matter how many men it has eaten.

When, in 1910, Jim Corbett was called upon by the Indian government to hunt down the Panar man-eater, he made straight for the area, which was not very far from where he lived at Naini Tal in northern India. Questions were being asked in the

British House of Commons about this man-eater, for it was picking over the same area as the infamous tigress of Champawat, and between the two cats 836 peasants were consumed—enough to populate a small village. Until Corbett made a move, no other hunter had heeded the government's appeal for somebody to kill the animal. The peasants' reaction, as always, was to barricade themselves in at night—and even during the day when they could. This often forced the Panar man-eater to tear down doors or burrow into grass roofs to get at the occupants.

On Corbett's first day in the leopard's hunting area he met a dazed and frightened young man who staggered from his lonely farmhouse and told the hunter how the night before the leopard had climbed over the balcony of his home, stolen into the bedroom, and grabbed his eighteen-year-old wife by the throat. The poor girl had screamed and then flung an arm around her husband, who held on to it and played tug-of-war with the cat. The leopard let go and retreated, enabling the husband to slam the door and hold it shut while the leopard, gaining new courage, tried all night to tear it down. The woman huddled in a corner until at dawn she lost consciousness. Even then the husband was too frightened to face the mile-long journey to his nearest neighbors.

Corbett saw the woman and noticed that her wounds—tooth punctures in the throat and four deep claw marks on her breast—were now septic. The room was windowless and filled with flies. The hunter was faced with the choice of going for medical aid, which was several hours away—for what he knew was a hopeless case—or hunting down the leopard while it was still in the area. He elected to wait for the return of the leopard, but he waited in vain. And the woman died.

The month was April and Corbett was forced to abandon the hunt until the following autumn. In September the leopard was still eating humans at an alarming rate. In the space of a few days it had eaten four men one after another in the same patch of brushwood. The villagers, not having a firearm, had allowed the leopard to finish its meals undisturbed.

Corbett decided to bait the leopard with a goat tied to a

stake while he balanced himself on a tree branch thirty yards away and waited for the cat to come. It came, but showed no interest in the bleating goat; instead it tried to climb the tree to get Corbett, who had taken the precaution of surrounding himself with thorns. He managed to shoot the leopard but succeeded only in wounding it. Later, with reckless courage, he sought it out by the light of a flaming torch and killed it.

Few of India's notorious man-eating leopards have shown signs of injury or old age, signs which might have explained why they suddenly took to man-eating. True, Kenneth Anderson's Gummalapur man-eater was found to be lame and his "Sangam panther" was old and decidedly worn in the tooth, but mostly they were in good health and their coats were silky.

It is significant that the Panar and the Rudraprayag man-eaters followed hard on the heels of epidemics of cholera and Spanish flu respectively.

The pilgrimage to Badrinath in the rugged, deeply forested Garhwal district of Uttar Pradesh in northern India is long and arduous. Annually for two thousand years, tens of thousands of Indians have climbed to the shrine of Badrinath in the temple on the shoulder of the 23,420-foot Mount Badrinath. The pilgrimage has always been a tough one and is only for the really devout. Between the years 1918 and 1926 it was particularly arduous because the sixty thousand pilgrims who trudged each year along the narrow, winding paths knew that watching their thin, straggling line was a leopard: the man-eater of Rudraprayag.

Occasionally the leopard would take a pilgrim but mainly it contented itself with coolly and calmly snatching villagers who were incautious enough to venture out of doors at night. If it grew impatient, which was often, and if no villagers showed themselves, it would then pad silently between the huts until it saw a door weak enough to tear down.

Occasionally the leopard found open windows or even groups of pilgrims sleeping in the open, and then it would swiftly and soundlessly sink its fangs into a sleeping victim and carry him away often without disturbing those who were sleeping next to him. For eight years, fifty thousand villagers lived in terror and observed a strict curfew that under any other conditions might have been intolerable. For eight years a fearful silence settled over the villages after sunset, for each inhabitant believed that the leopard was listening outside his door.

The government appealed to hunters and employed India's finest shikaris at elaborate fees, but the man-eater was too cunning and the men who hunted it became infected by the villagers' dumb terror. Even though there were something like four thousand licensed guns in the district the government granted three hundred more—just to aid the hunt. But although the footprints of the hunters in the deep, echoing ravines often obscured the pug marks of the man-eater, the Rudraprayag leopard carried on undaunted.

The government offered ten thousand rupees and the rent from two villages to the man bold enough to kill the cat but there were no takers. None—except Jim Corbett.

The hunter's first attempt failed. This was due mainly to Corbett's own ill health and partly to the superstition of the villagers, who by now believed that the leopard was supernatural. Then, on May 1, 1926, Corbett, using a goat as bait, shot the leopard dead. It was an anticlimax in a way. The official toll of the leopard was 125. There have been worse leopards before and since, but somehow none has ever achieved the same notoriety or instilled the same degree of terrorism as this one.

Other Cats

"I like little pussy, her coat is so warm,
And if I don't hurt her, she'll do me no harm . . ."
Nursery Rhyme

The jaguar has a fascinatingly evil reputation; it also has a large number of human kills to its credit—but it deserves neither. *Panthera onca,* the world's third biggest cat, is an enigma, but it can hardly be described as a great man-eater. It is a timid cat and should you meet one sliding through the greenery of a South American jungle you can be sure it will be the cat that will be the first to jump into the undergrowth and disappear. It has no truck with man.

Ernest Thompson Seton, in his 1925 book *Lives of Game Animals,* tells the story of the man-eater of the Rio Bravo—a jaguar whose career began the same day it ended, April 10, 1825, at the Convent of San Francisco, Santa Fe, New Mexico. A lay brother, having said his prayers, stepped from the chapel into the sacristy, through which he had to pass to reach the garden. The sacristy was a dark place for it was windowless. There was a short scream and a deep, rumbling growl. Then silence. A convent guard who heard the noise went to investigate. He walked into the gloomy room and found himself facing a jaguar. The cat killed him. Hearing the commotion, a group of monks came to investigate and the bolder of them stepped inside —he too was promptly killed.

A Senator Iriondo, a visitor to the convent, then took charge, called for volunteers, and decided to go behind the sacristy. They were hardly round the corner when they heard an awful cry and realized the jaguar had nipped into the garden to kill a fourth man and was now dragging him into the sacristy. Iriondo sensibly closed both doors to the sacristy and called in a hunter, who shot the animal through loopholes bored into the door.

Seton assures his readers that the authenticity of the story cannot be questioned, for his source was a United States government report of 1857. The section of the report dealing with the man-killer was written by Professor Spencer F. Baird, assistant secretary of the Smithsonian Institution, who claims to have taken it directly from the records of the Convent of San Francisco, Santa Fe. But the story is false. Whether the professor deliberately set out to hoodwink the government or whether he had trusted to hearsay is impossible to say now. Nonetheless, the Convent of San Francisco never existed and no lay brothers or monks worked in New Mexico during this period. As for Senator Iriondo, local historians found that he was fiction too.

Technically, it is just possible for a jaguar to be found in this region, which marks the absolute northern extremity of its range. According to Peter Matthiesen, the jaguar was once found as far north as the Red River in Arkansas, but today it can be considered virtually extinct in North America. Robert Froman, who did a great deal of work on the jaguar, records a solitary jaguar near Ruby, Arizona, in 1956 and another in the Patagonia Mountains east of Nogales in 1958 and yet another (a mother and two kittens) in the valley of the Sonora River just a few miles south of the Mexico-Arizona border in 1960. Froman believes the jaguar is expanding its number and its territory, an interesting opinion since its not very distant relative, the leopard, is doing equally well in Africa.

The jaguar looks like an overweight leopard at first glance and then one notices that, unlike the leopard, its peculiar rosette markings have blobs in the centers and that the skull is heftier and the cat less agile. All the same, the jaguar, if it but knew it, could be a more efficient man-eater than the leopard of India for it has far more strength and the added advantage of being at least 50 percent heavier. The biggest specimens come from the jungles of Brazil, the center of the largest jaguar population. Stragglers are found as far south as south Patagonia.

El tigre (as the Spanish-Americans call it) is undoubtedly the strongest swimmer of all the big cats and is quite at home in the water; in fact it is an expert fisher and uses this talent to supplement its varied diet. It is not a great tree climber. Although it will occasionally chase monkeys up trees, it will rarely if ever seek refuge in a tree if pursued by man. This speaks well for the cat's intelligence.

The disparity of opinions concerning the jaguar's character is amusing if unhelpful. Bernard Grzimek describes it as very dangerous while Alexander von Humboldt, like Grzimek a German and also a very accurate scientist, wrote: "There is no suggestion that it is ever a serious menace to man." Humboldt, who explored South America in the early 1800s, records an inquisitive jaguar loping up to an Indian child (a curious piece of behavior which crops up from time to time) and the child sending the jaguar rushing back to the jungle by giving it a swipe. Froman says the jaguar is "not very bold when faced by a man" but goes on to say that a "full-grown jaguar is more than a match for any man" and that he tries "to lead his pursuer into dense brush for the purpose of ambush." Theodore Roosevelt in *Animals of Central Brazil* describes the jaguar as "formidable" and says "they will charge men and sometimes become man-eaters." Dr. George G. Goodwin of the American Museum of Natural History adds his opinion: "The jaguar is the only American animal that becomes a man-eater by habit."

Froman himself, after sifting through a pile of literature on the animal, came to the conclusion that "it is very difficult to come by a first hand account." He offers one account from Roosevelt's *Through the Brazilian Wilderness,* which describes how an Argentinian was struck by a jaguar when he got in the way of the cat as it was foraging among some huts. The Argentinian was killed by the blow. The animal showed no further interest in its victim. This isolated incident gets us nowhere and yet it is the only mention Roosevelt ever made of a jaguar killing a man. It suggests that Roosevelt's descriptions of the jaguar's formidability were based upon hearsay. Phillip Welles, biologist in the Coronado National Memorial on the Arizona-Mexico border, says that during his travels he heard a lot about the jaguar's alleged man-eating habits but found no reliable accounts. He found that a lot of stories of man-eating in Mexico dated back to about 1911 when Mexico was torn by revolution and the territory to the north of Vera Cruz suffered badly. The biologist suggests that perhaps with all those bodies lying around the local jaguars might have acquired a taste for human flesh.

Azar, the South American writer, leaves us with an aggravatingly scant account of a jaguar which attacked a camp and took a dog, then came back for a Negro, then later an Indian, and then a Spaniard. Eduardo Barros Prado in *The Lure of the Amazon* claims to have rid the lower Aripuana area of a man-eating "tiger" (local name for the jaguar) "which had established a reign of terror."

Prado's jaguar, shot many years ago, was apparently one of a pair of man-eaters, for he mentions Jose Santiago, "a renowned tiger hunter . . . [who] met his death in the same district when he came face to face with a tiger which had been ravishing the locality." Prado does not say to what extent this dead jaguar had been man-eating. Santiago's body was found next to the jaguar's and it was assumed the trouble was over but then, "a few months later, a little boy, fishing by a stream, was attacked and in the next twenty days, six victims were dragged away by the man-eater, the last being a mother washing clothes near her home." The small son of the last victim led Prado to the jaguar's lair, where Prado was very nearly killed. He was camping near the lair at dawn when the jaguar sprang at him from a bank. Prado managed to get two bullets through the jaguar before it hit the ground. Two more shots into the crippled cat were enough. His shaken boy companion then turned his face to the sky and said in between sobs of relief, "You saw that, mother? I tell you, the gentleman has killed it. I tell you, he has killed it." The boy then said to Prado, "God bless you." Prado comments: ". . . this was an allusion to a belief widely held among Amazon people that the soul of a person devoured by a jaguar cannot rest in peace until the animal is destroyed." This suggests that man-eating, at least in this part of Brazil, is not infreqent. Earlier Prado and his assistants shot two troublesome jaguars in the same area. These jaguars had been worrying road gangs but what exactly they were up to Prado does not say.

In spite of Prado's experiences one wonders whether he was not carried away when he wrote: ". . . the tiger is the most deadly of all the living animals. He has plenty of intelligence, is utterly pitiless and his ferocious cunning is combined with complete fearlessness. A man has as much chance against a tiger as has a mouse against a cat, it is a menace that cannot be ignored."

Looking at the evidence—the evidence that can be relied upon—it seems clear that the jaguar does become a man-eater but very rarely. It does not appear to have produced any individuals which can compare with even minor man-eaters among the tigers, lions, and leopards of Africa and Asia.

Is it possible that there is another animal in the world that has as many English names as the puma? It is also called the cougar, mountain lion, panther, painter, catamount, silver lion, purple panther, red tiger, brown lion, deer tiger, Mexican lion, American lion, mountain screamer, cat-of-the-mountains, and Indian devil. In view of its silver, purple, red, and brown labels it is amusing that science has chosen to call it *Felis concolor*. But call it what you will, you are speaking of a very wise old cat. Like the African leopard it can live undetected near cities. According to newspaper reports, forty were sighted in a couple of years in the New York, New Jersey, Vermont, New Hampshire, Massachusetts areas. Others were found within thirty miles of downtown Boston and within thirty miles of the northeastern suburbs of Birmingham, Alabama. Although the puma has been exterminated in many areas it is still the most widely distributed large animal on the two American continents and is generally in no danger of disappearing. Pumas hunt from the snows of the Canadian north to the hot, dry plains of Patagonia and from the Pacific seaboard to the Atlantic. They have adapted themselves to the moist jungles of South America, to the arid desert areas in North America as well as to the freezing tundra areas of the Andes and the Rocky Mountains. Only the Asian tiger is as versatile in environment.

The puma weighs about the same as the African leopard—from 100 to 175 pounds—and can attain a length of nine feet. There was an enormous one of 276 pounds shot some years ago. The record length is 9½ feet.

As a man-killer the puma is not a serious menace at all. As a man-eater, apart from a report from the State of Washington in 1924 when it is supposed to have partly devoured a boy, it is innocent. Peter Matthiesen quotes John Lawson, surveyor-general of North Carolina in 1711, as saying that cougars were a greater menace than wolves but then, as wolves bite fewer people than do domestic dogs, this does not take us very far with the puma. Professor E. L. Jordan of Rutgers University claims that "there is no authentic record of an unprovoked attack on man." His statement is interesting in that it was made three years after seven-year-old Dominic Taylor of Vancouver was carried off and killed by a puma.

From all accounts this attack seems to have been unprovoked. Robert Froman reports that it happened on the northeast coast of Vancouver Island near the village of Kyuquot on June 12, 1949. Dominic was picnicking with his parents on the beach and had wandered a little way off when the puma, which had been lurking in the undergrowth, rushed out, seized him and carried him into the undergrowth. The boy's father raced to the rescue shouting at the top of his lungs. The puma dropped the boy, whose face and throat had been badly torn. Dominic died six hours later. There exists the possibility that the puma mistook the boy for some animal. In any event the attack was a freak and, according to Froman, is the only fatal attack in the province of British Columbia this century.

Froman in his 1957 book *The Nerve of Some Animals* states: "Our native lion is ranked as disgracefully harmless [in our folklore]. Even when cornered it supposedly permits itself to be shot and killed without trying to fight back. . . . on the other hand mountain lions have attacked humans viciously and in no distant, semi-legendary past. Since 1948 nine reports of lion attacks on humans have been accepted by competent authorities as apparently authentic—one each in New Brunswick and Idaho and seven in British Columbia."

The attack in New Brunswick which Froman mentions could have been due to the cougar's mistaking a man for an animal. It was in June 1948 in York County, just north of the Maine border. Alphonse Saulnier, who was working in a pulp factory near Fredericton, went down to the river to drink and was in a crouching position when a puma leapt on his back and sank its teeth into his shoulder. It clawed him badly as he kicked at it—and then it retreated.

Matthiesen is not quite accurate when he states: ". . . the documented instances of cougar attacks on human beings, in all the history of North America, can be counted on the fingers of one hand." In fact there have been something like three to four dozen incidents in this century, although most have been of a minor nature.

It might be argued that pumas used to be a great deal fiercer than they are now. An account of 1800 describes them as "fierce and ravenous in the extreme," although it does not state whether this attitude was toward man or their normal prey. Armand Denis records a tombstone in Pennsylvania dated 1751 on which is carved a puma and which indicates that a man named Philip Tanner was killed by a puma close by. But the fact that so few stories of pumas making attacks have been recorded in America indicates that the puma has always been a timid cat and has thus never constituted a threat to man in any part of its range.

Of all the wild cats, the cheetah is probably the most amiable. There is nothing on record to suggest it has ever killed a man, although, physically, it would be quite capable of doing so. There are, however, a few cases where it is supposed to have attacked human beings but never with any disastrous effects.

Denis describes the cheetah as "stupid, indifferent and stubborn" and "in regards to their tameness, there is much to be desired." My own experience with cheetah is that they are extremely affectionate and, unlike their cousin the leopard, are not given to dangerous bursts of temperament.

João Augusto Silva says that if a child "passes in a care-free run it is enough to awaken the cheetah's hunting instinct and it will shoot off, dangerously to give chase." This is perfectly true. A friend of mine used to amuse himself and his pet cheetah by running past it knowing that the cheetah would be unable to resist giving chase. The cheetah would lope up alongside him and trip him just as it would a buck but at no time did it ever show any signs of being dangerously excited.

Stevenson Hamilton, a man who, as warden of the Kruger National Park in South Africa, knew cheetah in the wild very well indeed, described them thus: ". . . though a ruthless hunter of his natural prey, a chita is perfectly inoffensive as regards man, and when wounded shows an entire lack of pugnacious qualities, presenting in this respect a most remarkable contrast to his relative the leopard. . . . he is gentle and timid . . . and docile in captivity."

The snow leopard (*Uncia uncia*) is a rare cat of the central Asian highlands. It is capable of putting up an impressive show of aggression when at bay and would, like any other cornered animal, attack, but I have come across no cases where it has.

The clouded leopard, also called the clouded tiger (*Neofelis nebulosa*) is found in Southeast Asia, Sumatra, Java, Borneo, and Formosa. Although it is capable of pulling down quite large animals, it usually sticks to small mammals and birds. There is a report of one attacking a group of men in Borneo without doing much damage. Generally it appears to be as unaggressive as it is handsome. The cat is about three feet in length not counting its long tail and weighs around forty pounds.

V
KILLER APES

The Chacma Baboon

Gorillas, Chimpanzees, and Monkeys

"One, with a soul and a mighty brain . . . the other, a wild caricature of ourselves."

"Animals Most Like Men" in The Children's Encyclopaedia

The Concise Oxford Dictionary is a sober little book. It squats there on my shelf, modest, quiet, unassuming; always willing to render what is asked of it, never more and rarely less. Except that is, when it gets to "Gorilla." Here it falls into the pit: "Gorilla, n. Large powerful ferocious. . . ." Large they are and powerful—but not ferocious. They are no more ferocious really than the Oxford Concise.

Since the Second World War our knowledge of the ways of gorillas has increased enormously, primarily from a realization that study of the giant primates enhances our understanding of ourselves. If man was never actually an ape, as we know it, then he is something disturbingly close, and looking through the whole gallery of modern apes and monkeys it is the gorilla that most nearly resembles man. For this reason a great number of false legends have grown up around this enormous parody of man.

Sir Richard Owen, the British biologist of the last century, succumbed to popular fiction when he wrote: "Negroes when stealing through the shade of the tropical forest become sometimes aware of the proximity of one of these frightfully formidable apes by the sudden disappearance of one of their companions, who is hoisted up into the tree, uttering, perhaps, a short, choking cry. In a few minutes he falls to the ground a strangled corpse." This image of a massive, chest-beating, man-strangling ape persists today in most people's minds. The gorilla, however, is a shy animal whose great displays of ferocity are a bluff. Unless he is driven beyond the point of endurance he will beat a noisy retreat when he sees man. If his family is endangered he will charge repeatedly, beating his midriff with his open hands, but if the gorilla's quarry makes a stand he will pull up short, scream with rage and frustration, and rush off into the undergrowth.

It is hard to believe that such a fierce-looking ape which can weigh up to five hundred pounds in the wilds (six hundred in zoos) can be so scared. Perhaps the old explorers were right that the gorilla used to be a dangerous animal but now, after sixty years of being persecuted by man and his gun, he has learned he cannot win. After all, according to recent observations of gorillas in the wild, they are scared of no other creatures, not even the leopard. But the validity of these opinions is doubtful. So few reliable accounts exist of any ferocious acts perpetuated by gorillas (apart from attacks by wounded or cornered animals), and it does not seem possible for the gorilla to have changed so quickly and so radically. His habitat shows that he is not a particularly adaptable animal. In fact, he is so specialized that he is without doubt doomed to extinction within a few generations.

Gorillas are found in only two parts of Africa: in the dense forests around the Cameroons in West Africa (this race is known as the lowland gorilla—*Gorilla gorilla gorilla*) and in the dense, wet forests in the region of the Albert National Park between the misty Virunga Volcanoes and the Ruwenzori Mountains in Central Africa. The Ruwenzori Mountains are the home of *Gorilla gorilla beringei,* the mountain gorilla. The differences between the two races are negligible although a thousand miles and probably many generations separate them.

In those brooding and damp mountain forests where the last mountain gorillas feed upon bamboo shoots and other vegetation, a local game guide named Rubin has survived three terrifying experiences with gorillas. It is an indication of the gorilla's geniality that this man, who for years has been in almost daily contact with gorillas, has experienced only three troublesome incidents. On one occasion he was following up the blood spoor of a wounded gorilla—the gorillas had been fighting a rival on and off for several days—when he suddenly found himself facing the animal. The enormous silverback (as old bulls are called) came straight for him. Now Rubin knew the rules as well as any man: one must not run from a charging gorilla but turn and face him, then *he* will run. At the same time Rubin could see that this gorilla was so badly wounded that he could not be counted upon to remember his side of the rules. So the African compromised: he began to dance wildly and loudly. So astonished was the gorilla that it pulled up and stood rooted to the spot. Then it turned and fled. On another occasion Rubin was watching from very close quarters a female gorilla who had been spurned by the local patriarch and was in a towering rage. She suddenly saw Rubin and charged up to him. Rubin, taken completely by surprise, struck her with his panga. The female bit him. Rubin then kicked her in the stomach. The gorilla broke off the fight and dived into the undergrowth. The same female threatened Rubin again two days later but the African shouted right back and the frustrated ape bowed out.

Carl Akeley, perhaps the best friend the gorillas ever had (for it was he who persuaded King Albert of Belgium to proclaim their area a game reserve) once said: "I believe that a white man who will allow a gorilla to get within ten feet of him without shooting is a plain darn fool." But George Schaller, who spent practically two years living in gorilla territory with the purpose of studying them, never carried a gun at first. His wife, Kay, later prevailed upon him, and he ended up carrying a harmless starting gun but never had cause to use it. Instead, he found his subjects "reserved and shy." Schaller says that on the rare occasions when gorillas do attack men they tend to bite and run. By all accounts bad maulings are very rare and local mission-hospital records indicate that of the two or three casualties a year none so far have been fatal. In one case a gorilla seized an African (who had wounded it during a wild and boisterous hunt) by the knee and ankle and tore away his calf.

Fed G. Merfield, a hunter in West Africa, found the race of gorillas there much the same as Schaller's gorillas: secretive and extremely nervous. These gorillas, protected since the 1930s, are still hunted by local tribesmen in the most ruthless manner. Merfield has described some appalling tribal hunts in which gorillas are trapped and speared until they resemble porcupines. Even then attacks are not common. In fact an African peasant who is wounded by a gorilla is ridiculed when he arrives back at the village because it is assumed he turned and fled instead of standing his ground. Merfield, who put this theory to the test many times, says: "Of course there are exceptions to this. If you happen to tread on a gorilla's toes in the forest, you can expect to be torn to pieces, but even then the gorilla will be more concerned with getting away than with killing you. His action will be to sweep you aside with his powerful arms and hands." This actually happened to Merfield.

Merfield claims that female gorillas "are completely harmless." He describes how native hunters, having killed the bull, will gather around the female and beat her over the head with sticks "and it is most pitiful to see them putting their arms over their heads to ward off the blows, making no attempt at retaliation."

In captivity the gorilla is a very different animal and it can, in adult life, be dangerous even to a keeper who has known it well. Captivity tends to have a bad effect upon a gorilla, and almost all of them die within a short while. Unlike its close relative the chimpanzee, which settles down well in captivity, the gorilla becomes morose and may refuse his food to the point of starvation.

The extroverted chimpanzee, which roams a wide area of Africa's rain forests, is shy but relatively amiable in the presence of man in the wilds and apart from an astonishing, although not necessarily unique, case of man-eating it cannot be counted as a dangerous animal.

In March 1957 in the Kasulu district of Tanzania an African woman carrying a baby was attacked on a forest path by a large chimpanzee. The incident happened near the shore of Lake Tanganyika north of Kigoma. The inquest papers carry the following evidence from the African woman: she was carrying the baby on her back when "suddenly from the bush came a chimpanzee. We were in the bush and the village was far. I was tying up my faggots. I ran away and the chimpanzee hit me twice. He was about four feet tall. I fell down. Then it caught the child who was on my back. I made a great deal of noise and other women came. Then we saw the chimpanzee eating the child's ears, feet, hands and head." According to the medical evidence, there were five depressed fractures of the skull caused by teeth, the scalp was missing, and so were the hands and half of one foot. The coroner ruled the death accidental and found that the child had been "eaten by some animal."

I say that this incident might not be unique because there is growing evidence to suggest that the chimpanzee does eat flesh. (This is unlike the gorilla, which, in the wilds at least, does not appear to eat meat at all). Jane Goodall watched chimpanzees in Tanzania eating the flesh of monkeys which they had first killed. This local sub-species is *Pan troglodytus schweinfurthi*, which is known as the long-haired or the eastern chimpanzee, and which is a particularly popular one in zoos and circuses. It grows to almost the size of an average man in weight, if not in height, and its diet comprises fruit, plant shoots, and occasionally eggs.

The chimpanzee in captivity is usually amiable and, when young, can be most endearing and show a real affection for people. When they get to ten years or so one can expect fits of bad temper. Elderly male chimps of fifteen years and more can be very dangerous indeed.

Dick Chipperfield, Sr., says that of all the animals in his circus (tigers, lions, leopards, polar bears, and Himalayan

bears) he regards his twenty-year-old chimpanzee Charlie
(caught in Sierra Leone) as the most dangerous. "I am only
just the boss, as far as Charlie is concerned," he told me once.

In the monkey world (comprising all primates except those
classed as anthropoid apes) the baboons are the only potential
man-killers but, by and large, even the biggest of them have
the intelligence to give man a wide berth. In fact "man" is the
operative word, for some baboons—notably South Africa's
chacma baboons—are much less afraid of women and will even
approach them making threatening gestures. This same species,
Chaeropithecus ursinus, has also learned to distinguish between
gun-carrying men and unarmed men and will allow the latter
to approach much closer than the former.

There are several cases recorded each year in South Africa
of baboons attacking humans but in almost every case the culprits are pet baboons. One such case occurred in a street in
Brakpan (a town near Johannesburg) on April 10, 1964, when
an eight-year-old escaped chacma grabbed a one-year-old child
from its pram, sank its fangs into the child's brain, and then
dropped the dead child on the pavement a few yards from its
horrified mother. Later the Transvaal Provincial Council declared
baboon pets illegal. In that year South African newspapers
carried half a dozen reports of people, mainly children, being
bitten by baboons, in one or two cases severely.

Attacks in the wild are very rare. My own experience of
chacma baboons is that they are probably the least nervous of
Africa's various baboon species. They will, nevertheless, retreat
if an unarmed man approaches to within, say, fifty yards. If a
man is carrying a gun they will begin running when he is still
as much as three to four hundred yards away. I have often tried
stalking them but I have never been able to escape detection
by the scouts which they post around the troop. These scouts,
usually big, old males, frequently bark as one approaches. For
somebody who is not used to these animals—they are quite
large, being about three and a half feet when standing—the
sight of their fangs, which are every bit as impressive as a

leopard's, can be disquieting. But apart from barking, they rarely make any threat.

An interesting police report appeared in the South African press on March 11, 1963, claiming that a large male baboon had been shot by the police in the Cullinan district after Africans had complained that it had thrown stones at them and repeatedly threatened them over a period of two years. The police discovered that this same baboon had, two years before they shot it, chased two African girls aged five and thirteen over a hundred-foot cliff to their deaths. On September 9, 1964, in Sutherland in the Cape Province it was reported that a laborer named Fred Visagie had practically all his clothes torn off by a big baboon, which attacked him on the farm Rooiwal while he was looking after some sheep. The man was badly bruised and scratched. He claimed that as the baboon was attacking him the rest of the troop gathered around to watch. Two months later in Fish Hoek on the Cape coast a pair of baboons terrorized a mother and her two children by trying to break into their home. The mother told the police, who shot both animals in the garden, that the baboons had been banging on the doors and windows. I wondered after reading the report whether the mother had not misinterpreted the baboons' intentions, because the incident was reminiscent of an episode in Eugene Marais's book *My Friends the Baboons*. Marais, who lived for three years in a hut among a colony of baboons in the Waterberg Mountains in the Transvaal, described how one night the baboon leaders came down and began banging at the windows. He saw their ugly faces at the windows but he knew them all individually and he knew them well enough to realize that something was radically wrong. He went outside and they made off up the mountain, looking back and waiting to see if he was following. Marais did follow and they led him to the rocky outcrop where they slept at night and there, by the light of his lantern, Marais could sense something was wrong. He soon discovered what it was: eight babies had died. He later carried their bodies down the mountainside to his hut and the mothers followed in silence. He took the corpses inside, but it was clear that they were quite dead and that they had died for want of a good diet (there had been a famine). For hours the mothers waited, but when they saw Marais could do nothing they turned and filed back up the mountain wailing mournfully.

The yellow baboon, which is found from Rhodesia north into Central Africa, has a similar record of sporadic attacks upon humans and P. J. Pretorius tells of a large, solitary male which took up man-eating. Baboons are becoming more and more inclined to eat flesh in many parts of their range and have been seen eating small antelope, which they had apparently killed. Meat-eating baboons were first observed at the beginning of the century and the phenomenon is causing a great deal of interest among anthropologists for at one time dawn man must have gone through a similar adjustment of his diet. Pretorius's man-eater operated at the side of a swamp in the Ruvuma River area of Tanzania and apparently rushed out at passing Africans (if they were alone) and disemboweled them with its fangs. It then would break open the skull by using its teeth and eat the brains, leaving the rest of the body. Africans were so terrified of the animal that in the end they abandoned their village.

The biggest of Africa's baboons is the mandrill (*Mandrillus sphinx*), which has a face as fierce as it is colorful. The face is sky-blue, red, and orange. But apart from one or two attacks by rogue males they live in fear of man and are extremely difficult to approach.

VI
KILLER REPTILES

John Pitts

The Nile Crocodile

The Crocodile

"A cruel crafty crocodile."
Edmund Spenser

In 1957 I was in the Black Umbulozi River area of Swaziland when I met a missionary doctor named Samuel Hynd who was looking for the mother of a six-year-old girl named Busisiwe Magagula. The mother had disappeared into the bush to mourn Busisiwe's death. Hynd told me that the child had been seized by a crocodile while she was playing in the Black Umbulozi and was immediately dragged under the water. The mother, hearing other children shrieking, rushed down to the river, waded in, and then saw the child holding on to the reeds on the far bank. She grasped the girl's hands and only then noticed that the crocodile was still fastened on to the leg. A grim tug-of-war developed between the mother and the reptile. Suddenly the crocodile spun over in the water and tore off the child's leg just above the knee. The child's father, fortunately an ex-South African Police constable, trained in first aid, tied a tourniquet with his old police belt and began to carry the child, at a trot, in the direction of the mission hospital at Manzini (then Bremersdorp), which was sixty miles away. At some stage of the journey a farmer found him stumbling along and took the child to the hospital. She appeared to be dead. Hynd told me, "I saw her carried in and it seemed certain that she would die of either shock, loss of blood, or infection. I have seen too many crocodile victims to have any illusions. But miraculously Busisiwe survived. In fact within a week she was walking on crutches and we found her the most intelligent and cheerful child."

Eventually they found the mother, who refused to believe the child was alive until she saw her struggling on crutches from the mission station to greet her.

When the story of Busisiwe appeared in the Johannesburg newspaper *The Star* it brought in large donations of money for the child's education, because most people realized that her marriage prospects would be nil and therefore she would need a career. An artificial limb manufacturer guaranteed her a new leg for the rest of her life. He went one further and every time she comes up to Johannesburg he buys her a new pair of shoes and socks and spends several hours matching the skin color of the artificial leg to the girl's other leg. Today Busisiwe Magagula is a very active mission worker and, at the age of sixteen, has a number of important responsibilities at the Manzini mission.

This story had a happy ending, but there are many more that have not. It is difficult to be rational about the crocodile. It *is* a hideous beast, but it can't help that. It is also a cunning and savage killer but then all predatory animals have to be both of these things. The one thing that can be said in favor of the crocodile is that it is at least one better than man: it kills neither for "sport" nor for revenge. Its actions are entirely defensible biologically and it plays a vital part in maintaining a balance between the predatory barbel (catfish), which is its staple diet, and the all-important fresh-water fish on which millions of Africans depend for protein.

Winston Churchill wrote when describing a Nile journey in which he saw crocodiles close up for the first time: "I avow, with what regrets may be necessary, an active hatred of these brutes and a desire to kill them." The old hunter Foran got completely carried away and wrote: "The loathsome and hideous crocodile . . . no one should ever hesitate about killing crocodiles wherever a chance offers, because it is either already a killer many times over or a future destroyer of life." Later, when he was discussing how he took pot shots at them from a Nile steamer: "I felt neither shame nor remorse. The savage and cunning killers ought to be shot exactly as one would a mad dog."

Crocodylus niloticus has been in the world 170 million years and, in spite of the tremendous amount of crocodile hunting that has been going on in Africa recently, will be with us for a long time yet—probably after many of the larger animals of Africa have become extinct (although one authority claims that the crocodile is in immediate danger of extinction). The formidable crocodile is an evolutionary triumph which has the advantage of not being particularly edible although it is eaten in some parts of Africa, of being able to live on land, on the water, or under the water, and of being able to eat almost whatever it fancies. It is armor-plated and its grip is almost unbreakable. Where most animals lose their second teeth and are thus unable to pursue their normal diet, grow weak and die, the crocodile's unique dentition—it has self-renewing teeth—allows it to follow its normal diet for years. Some zoologists suspect it might live at least as long as man. Its strength is amazing and full-grown crocodiles weighing, say, three quarters of a ton have been known to pull rhinoceroses and buffaloes into the water and drown them.

One of the most controversial features of the crocodile is its length. Cherry Kearton did not help the argument much when he claimed in his book *In the Land of the Lion* to have seen a 27-footer on the Semliki. Now a 16-foot crocodile in the Lourenço Marques museum stands as high as a man's chest, which means Kearton's crocodile would have been higher than a horse, had the girth of an elephant and weighed just as much. Colonel Pitman, who shot one hundred of the biggest crocodiles he could find on Lake Victoria, found only three more than 15 feet and they were only just over that measurement. Jack Bousfield of Lake Rukwa in Tanzania, who had a hand in killing forty-five thousand of them, says he never saw one over 18 feet. The biggest shot "officially" was on the Semliki River by the Uganda Game Department in 1953—it measured 19 feet 9 inches.

According to people who keep crocodiles as pets or breed them on crocodile farms, one can become quite attached to them. Of course, there is always the danger that they might become attached to you. I must confess to feeling the way most people feel about these reptiles: they have an infinite air of evil about them. The shark's smile is at least benign and it does hide its razor-sharp teeth but the lipless smile of the crocodile hides nothing: he flashes all his crooked teeth. He has an upper jaw which Robert Ruark admirably described as being like a spiked manhole cover and his tremendous mouth is nothing more than a highly efficient, rustless steel trap. Even my two small daughters, when they saw their first crocodile sliding across a gloomy African river with one casual flick of its oarlike tail, were struck dumb with awe, for here was something they seemed to instinctively recognize as a man-eater.

A few years ago I was sitting beside a narrow channel in the St. Lucia district of Zululand talking to the engineer in charge of reclaiming the once world-famous big-game fishing ground. Recent floods had caused the Umfolozi River to burst its banks and had pitched hundreds of crocodiles into the St. Lucia swamps and the channels were alive with them. As we talked, at least a dozen crocodiles lined up forty feet away under the shadow of the opposite bank. Some Africans were standing near the edge of the bank watching them and I mentioned to the engineer the possible danger of one of them being flicked off by a crocodile's tail. "It's impossible," said the engineer. "How can a crocodile tilt its body or lift its tail up high enough to swish somebody off a riverbank?" It is odd how one can believe something for years and then sudenly realize it is impossible. The engineer was right, but the belief that crocodiles can knock people off banks is considered gospel right out of Africa. According to Pete Wessels, they can rocket out of the water onto land and snatch an animal standing on the bank. "I have seen them land on rocks six feet and more above the surface," he says.

When we got up to go the twelve floating crocodiles sank out of sight without leaving so much as a ripple. An hour or so later one of the Zulus leaped unconcernedly into the river to push a pontoon away from the bank. In two seconds there was a flurry of water and he was gone. Only one crocodile appeared to be involved although several others were around. Since crocodiles are inexplicably good-mannered over food and seldom fight each other, the crocodile that had seized the man was allowed to part in peace with his struggling victim, who twice broke the surface and waved frantically to his horrified companions. He was never seen again. Incredibly, the next day, I watched Africans swimming across the same channel.

The African attitude toward crocodiles is an odd one. In some parts they swim and bathe within sight of crocodiles, claiming that as there are plenty of barbel for the crocodiles to eat they will not be interested in people. Lieutenant Colonel Stevenson Hamilton tells of a crocodile at Lujenda in Portuguese West Africa which took an average of two tribesmen a month. The colonel asked the chief why, as he had a muzzle-loader, he did not shoot it. "What!" said the chief, aghast at the thought. "We can't afford to waste our gunpowder on crocodiles." Africans are fatalistic about wading across rivers, even within hours of a man's being taken by a crocodile.

There are undoubtedly some crocodiles which actually acquire a preference for man-eating and possibly the worst case on record was on the Kihange River in Central Africa, where over the years four hundred people had been taken by a fifteen-foot-three-inch crocodile later shot by Wessels. Another crocodile, this time on the Zambesi, killed about three hundred people. Wessels recalls a sobering conversation with a missionary-school teacher who had lost one of his pupils to a crocodile: "Pete," he said, "he was the seventh this term."

Most crocodiles would eat a man if they were hungry and it would probably be fair to say that if a man persistently bathes in a crocodile-infested river his chances of being taken are greater than his chances of getting away with it. In populous areas crocodile hunters frequently find human remains or trinkets such as beads or bangles inside crocodiles. Most of those killed are women and children, because women with their children spend a great deal of time down at rivers drawing water or washing clothes. A lot of women are seized by the wrists as they bend over the water.

A story that one hears throughout Africa, and that might possibly have happened, is about a woman who, on being seized by the wrists, immediately fainted. The crocodile dragged her under the surface and then pushed her into its cavern underneath a riverbank. Although the entrance to a crocodile's cave is under water the actual cave is above the water line. The woman regained consciousness and seized with terror she pummeled at the roof until it collapsed. When she appeared in her kraal that evening the villagers were already mourning her death. It was some time before they accepted the fact that it was not her ghost that they were seeing.

One thing that might explain why crocodiles are worse in some areas than in others is that most African tribes do not believe in burial and some prefer to throw their dead in rivers. Some even throw their hopelessly sick in as well. A few still stick to the (now mostly illegal) custom of disposing of deformed babies or the weaker of twins by feeding them to the crocodiles. The number of Africans eaten annually by crocodiles has been a controversial topic for years and the highest at which the toll has been put is 20,000 by T. Murray Smith. Smith's figure is doubtless too high and although the toll in the days of which he writes—a generation ago—must have been great, I doubt it would have gone over 3000 or at the most 4000 a year. There were more crocodiles around then, Africans possessed fewer guns, and medical facilities were bad, to say the least. Today the annual toll of people killed by crocodiles throughout Africa—going by the numbers in those areas I know—must be at least 1000. Probably at least that number are injured.

Out of the water the crocodile is as nervous around man as an antelope is and as one walks along an African river one sees the dark shapes snaking down the banks to splash noisily into the river. Once in the water their attitude changes and they float, watching with their eye ridges and nostrils just above the surface; and not a ripple betrays them. Occasionally, especially toward dusk or very early in the morning when the light is bad, people step on crocodiles in the shallows, mistaking them for logs, but mostly in cases such as this it is the crocodile which gets the bigger fright. Alexander Barnes, author of *The Wonderland of the Eastern Congo*, stepped on to a crocodile at Fort Jameson. The reptile seized his foot but Barnes managed to throw an arm around a tree and hang on for his life. With his free hand he pointed his rifle down at the crocodile and pulled the trigger. The stricken crocodile let go and Barnes, after rubbing salt into his wounds to keep them from going septic, walked to the hospital. In recalling the incident, Murray Smith mentions that Barnes was later killed in another stream: a stream of traffic in New York.

There is yet another controversial aspect regarding crocodiles: some authorities claim they will never eat fresh meat. In fact they do. They seem to prefer fresh meat to old meat, and certainly when they catch a small antelope or dog they lose no time in throwing their heads back and letting it slide piecemeal down their throats. The head action of a feeding crocodile is necessary because they have no conventional tongue nor are they able to chew food. If they come across a carcass too big to swallow they will queue at it in a most orderly fashion. When each crocodile's turn comes, it will grip a piece of the carcass and spin horizontally until it is twisted off. If the carcass is too big for this type of treatment, the crocodile will leave it until it is putrefied. Some authorities claim they can swallow only small pieces of meat but this is not true: I have seen several instances where sizable dogs and even half the torso of a youth were removed from a crocodile's stomach. One of the most horrifying collections of stomach contents taken from an African crocodile was in East Africa and comprised the following: several lengthy porcupine quills, eleven heavy brass arm rings, three wire armlets, some wire anklets, a necklace, fourteen human arm and leg bones, three human spinal columns, a length of fiber—the type used for tying up faggots—and eighteen stones. One often finds smooth stones in the animals' bellies and nobody is sure whether they are swallowed to aid digestion, to act as ballast, or merely because crocodiles feed on river-bottom fish such as barbel and scoop up stones accidentally.

There are twenty-three species of crocodilian in the world. A third of these species are found in America, one in Cuba, four in Africa, six in Asia, and two in Australia. All but two are restricted to the tropics. Nearly all of them are well disposed toward man or are incapable of doing him any great damage. There is certainly none that is comparable with *Crocodylus niloticus* as a man-eater although the salt-water crocodile, *Crocodylus porosus,* has been known to kill and eat humans. The salt-water crocodile appears to be larger than the African crocodile but apart from that there is little physical difference. As a long-distance swimmer it outdistances all the others, and a solitary wanderer turned up on Cocos-Keeling Island a few years back. One or two have pitched up on the beaches of Fiji and even the New Hebrides, which suggests they swam many hundreds of miles. They are most numerous around the East Indies and a number of attacks have been recorded at sea off the Malay Archipelago.

The mugger or marsh crocodile of India (*Crocodylus palustris*) was said to eat an average of 250 Indians annually in prewar years but is so rare today that it is protected and man-eating must be practically nonexistent. The American crocodile (*Crocodylus acutus*), according to most authorities the biggest crocodile in the world, growing to twenty feet and more in extreme cases, is accused of man-eating throughout its range, which is chiefly Central America. There seem to be no instances, though, of it man-eating in the extreme north of its range, where it enters the United States. Most people who know the animal treat it as intelligent people treat the Nile crocodile—with a great deal of respect. It is almost as sea-loving as its near relative the salt-water crocodile.

The Orinoco crocodile (*Crocodylus intermedius*) is also one of the larger of the crocodylidae and Alexander von Humboldt claimed they grew up to twenty-four feet. These long-snouted crocodiles are not particularly noted for man-killing or man-eating, although they would be just as capable as the African crocodile.

Australia's little crocodile, *Crocodylus johnsoni,* is not considered dangerous.

The second member of the order Crocodilia is the gavial or gharial (*Gavialis gangeticus*), which has long, slender jaws, and although a number of those shot in the Ganges Valley have bangles and other human trinkets inside them it is almost certainly because they eat bodies which they find floating down this river. Although gavials grow quite big—thirteen feet full-grown —they are not considered particularly dangerous and I don't think there is a charge of man-eating against them.

Alligators have always been viewed with uneasiness by people and some writers have been completely carried away by their appearance. William Bartram in 1791 wrote: "Behold him rushing forth from the flags and reeds. His enormous body swells. His plaited tail, brandished high, floats upon the lake. The waters like a cataract descend from his open jaws. Clouds of smoke issue from his dilated nostrils. The earth trembles with his thunder." Although the crocodile has a voice it can make neither smoke nor thunder.

There are very few instances of aggressive behavior of alligators toward man, although E. A. McIlhenny was attacked by a twelve-footer when he was in a boat. He was flung from the boat and was forced to shoot the animal. Pope says that females will defend their nests most vigorously and the chance of a man being attacked through this set of circumstances is about one in a hundred.

The alligator is protected and is now mostly confined to Florida, but it used to be quite common in Arkansas, Texas, Carolina's coastal region, and the Mississippi Valley. In fact its scientific name is *Alligator mississippiensis.*

The most obvious difference between a crocodile and an alligator is in dentition. The true crocodile has teeth that form a single, exposed row when the mouth is closed, each of the lower teeth interlocking between each of the upper teeth. The fourth tooth in all crocodiles is prominent and nestled in a notch on the outside of its upper jaw. The alligator's lower teeth fit inside the upper teeth and are hidden by them. The fourth tooth is long, like the crocodile's, but fits into a slot inside the alligator's mouth.

A few years ago the naked and mutilated body of a small boy was found in a Florida swamp and there seemed little doubt that the mutilations were due to alligators. This was all the excuse that was needed for a frenzied campaign against alligators, a campaign from which the alligator population has never really recovered. Human revenge is normally swift, usually disastrous, and rarely rational, and this episode was no exception. Who stripped the little boy? The alligators? It seems more probable that the boy was thrown into the swamps dead or that he stripped for a swim and drowned and was then bitten by the alligators, if indeed the marks on his body were those of the reptiles.

The Gila Monster

"It is the green-ey'd monster which doth mock
The meat it feeds on."

William Shakespeare

Denis Groves, former curator of the Transvaal Snake Park near Johannsburg, telephoned me one morning and said: "You can write this down. Gila monsters haven't got a fatal bite."

"How do you know?" I asked him.

"I was bitten by one yesterday afternoon," he said. He explained that he was holding the monster showing it to the director of the Snake Park, B. J. Keyter. As he went to put it back it lunged at his thumb and bit it badly but fortunately did not get a grip. Groves, who had already been bitten by a dozen venomous snakes in the previous two years, did not want to tell Keyter "because every time I get bitten by something poisonous the boss gets ill worrying about it." So he thrust his hand in his pocket until the director had left. Fifteen minutes later he was able to wash and dress the wound but otherwise he left it untreated and went home at the usual hour. "The night was hell," said Groves. "I thought I would die. My thumb swelled up to a tremendous size and my whole arm was agonizingly painful." The following morning his thumb had gone down but was still tender. Later that day it returned to normal.

But it proved nothing really. One thing was clear: Groves was extremely lucky that the gila had let go so quickly. But small though the injection of venom had been, Groves still suffered badly. What would have happened if the gila monster had really got a hold—with that bulldog grip that often only death will make it release?

Heloderma suspectum and its close relative, the Mexican beaded lizard, *Heloderma horridum,* are a controversial pair. They are the only known venomous lizards on earth. The former is a thickset, bull-headed, rough-skinned, rather stumpy looking character which grows to about twenty inches. It has a mottled skin and a barred tail, which is short and fat. From the side view it looks not unlike a sawed-off iguana. The Mexican "monster" is more slender and grows at least ten inches longer. Its body is mottled and rough-skinned and resembles a monitor lizard's but its head is heavy, squat, and covered with shiny "warts."

Some research on gila monster bites was done by Weldon D. Woodson and was first published in *True West* (December 1964) under the title "Can the Gila Monster Kill?" Woodson's quest turned up 183 cases of gila bites in one hundred years of which 32 were said to have been fatal. Woodson then chose ten random case histories for his article and left the reader to draw a conclusion. The conclusion could only be that the gila monster *can* kill and *has* killed people.

A number of people who ridicule the idea of the gila monster's being lethal to man claim that those who died after its bite were all alcoholic or sick. Curiously it seems a great many of them were, but then if an alcoholic is knocked down by a bus did the bus kill him or did the alcohol?

The gila's lower jaw is equipped with venom sacs which, when the teeth have punctured the victim's skin, exudes a neurotoxin similar in action to the venom of a cobra. People bitten nowadays, provided they have quick medical treatment, including antiserum, should recover rapidly. There appear to be no recent incidents of fatal gila monster bites. Many of the early cases unearthed by Woodson, however, point to an appalling lack of action on the part of doctors. In all fatal cases in Woodson's report, the gila got a good grip—a grip that either had to be wrenched open with instruments such as pliers or that could only be released when the animal was killed.

In the Philadelphia *Times* of June 22, 1893, the following article appeared about an incident in Florence, Arizona. "Richard

M. Farthingay, a tourist from Minneapolis, returned here last evening with the remains of Arthur James, who had accompanied him on the journey, and who died the day before from the bite of a gila monster." It went on to describe how the two men had decided to sleep in a cave because it was raining. As they slept Farthingay felt something run across him. It was dawn and just light enough for him to see a gila monster making for his friend. The lizard climbed on to James's chest. Not knowing what it was, the half-asleep man grabbed it by the tail whereupon the monster buried its teeth in his wrist. "Mr. Farthingay hastened at once to his friend's relief and endeavored to pull it off, [but] it held on like the grim death it was.

"Then though fearing to strike the man instead of the reptile, he seized his gun and fired the contents into the creature's body. The monster let go its grip on Mr. James and made an effort to reach this new antagonist, but a second volley tore its head from its body." By now James had fainted. Farthingay revived him and the man complained of a terrible pain in the arm, which was beginning to swell to three times its normal size. Within half an hour the limb was discolored, showing that blood vessels were ruptured. Some of the skin was nearly black. James pleaded with his friend not to leave him and Farthingay decided to await developments. In a short time the victim became delirious and then raving. Finally he rushed shrieking from the cave to the river to quench what was apparently an overwhelming thirst. But James never drank from the river. He collapsed at the edge of it and died.

"When he reached him it was to find life entirely extinct, James lying with his head in the water and his own teeth fixed in the swollen, gangrened arm."

The Snake

"A running brook of horror . . ."
John Ruskin

The twelve-foot black mamba was clearly angry and every time somebody came near its glass pen it reared and struck at the glass. B. J. Keyter, director of the Transvaal Serpentarium, was worried that the snake might damage its fangs and then it would be useless to "milk" for venom—and as he had just paid nearly $1.50 a foot for this particularly deadly reptile the possibility was disturbing. He decided to occupy the snake's attention by throwing a white mouse into its pen for it to eat. He did just this. The little mouse immediately ran up to the rearing snake and bit it firmly in the side. Then it retired to a neutral corner and began cleaning its whiskers. Two days later the mamba died of infection from the mouse bite. "That's the trouble with snakes," Keyter told me. "They're such delicate things."

Snakes are strange creatures. The majority of species are incapable of exerting themselves to any extent, for only the left lung has survived the evolutionary processes which have been necessary to make snakes what they are—just one long, narrow rib cage with a head at the end. This remaining lung is so elongated that the rear end is almost nonfunctional. Nor have these animals any antibodies in their blood, not in the normal sense anyway, and thus a tick bite, if it is infected, can kill a highly venomous snake.

Of the world's 3000 kinds of snakes, only 300 have a well-developed venom apparatus, and of these, 50 are sea snakes, which very rarely bother man. Only about 150 species of snakes can be regarded as really dangerous to man. Some are exceptionally dangerous; one of them, the king cobra, is considered by many zoologists to be the most lethal animal in the world.

There are seven major groups of snakes in the world, two of which do not concern us, for they are absolutely harmless— the blind snakes and worm snakes. The biggest family is Colubridae, which comprises the "ordinary snakes" such as the grass snakes and other harmless species. One or two of this family carry a mild venom which is less harmful to humans than bee stings. There is the closely related but far more troublesome family Elapidae, which comprises such toxic species as cobras, mambas, and coral snakes. The Viperidae comprise the adders and rattlesnakes. The fifty species of sea snakes, which, apart from their specially adapted oarlike tail, closely resemble cobras in appearance and in venom, belong to the family Hydrophilidae. Finally we have the massive pythons and boas, which some scientists lump under Boidae and others separate into Boidae and Pythonidae.

It is impossible to estimate within two or three thousand or so the annual death rate caused by snakes but in June 1963 the World Health Organization (WHO) issued the following statement: "Forty thousand people are killed by snake bite every year according to conservative estimates made by WHO. Most of the deaths, perhaps 70 percent, occur in Asia." The figure may be short of the mark. Parts of Burma have the world's highest proportion of fatal snake bite to population—30 deaths per thousand people, according to V.F.M. FitzSimons. The country with the highest number of deaths, however, is undoubtedly India, which loses perhaps 20,000 people a year. The United States, which has the rattlesnakes and one or two other dangerous snakes to contend with, loses just over 100 people. In Africa, where few written records exist for this sort of thing, the annual toll must be around 7000. This figure might seem surprisingly low when one considers that Africans have to contend with three times as many species of cobra as

Asia and some thoroughly dangerous adders. Undoubtedly if Africa's human population were as concentrated as India's the death rate would be at least as high.

A Nairobi safari firm instructs its employees that if a client is bitten by a venomous snake while on safari they must assure the client the snake was harmless. They are also told to say that the injection of antivenin is a precautionary measure and, presumably, if they have to apply a tourniquet, they pass it off as a practical joke. It sounds unnecessary, and yet the idea is sound, for the more a patient relaxes the less the heart pumps and the longer it takes for the venom to travel through the body. Apart from that it is quite possible for a person to die of shock or autosuggestion after being bitten by a perfectly harmless snake.

Quite a number of deaths from harmless as well as toxic snakes are caused by the victim's drinking alcohol because of a widespread belief that it helps save lives. In reality, alcohol stimulates the heart and quickens the action of the venom. I have come across a few cases where men have swilled a bottle of brandy after a snake bite and have actually died of alcohol poisoning.

It is alarming sometimes to read the advice given by big-game hunters on snake bite. Murray Smith, in his book *The Nature of the Beast,* claims that puff adder bite wounds should be criss-crossed with a sharp knife or razor blade, that a tourniquet should be tied above and below the punctures, and that the wound should be irrigated with permanganate. Today all three recommendations are considered both useless and dangerous. Permanganate has been carried all over Africa by hunters and explorers but it has long been proven worse than useless. Cutting a snake-bite wound only helps the poison enter the subcutaneous tissues more efficiently and tying a tourniquet after an adder bite usually causes such terrible damage to the affected limb that it has to be amputated. South African herpetologists recommend neither practice for the treatment of snake bite. I have hospital records of fatal cases where doctors and nurses have actually precipitated death through the most alarming ignorance of the nature of snake venom.

There are two basic types of snake venom. The cobras carry a highly efficient neurotoxin which attacks the nerves and can cause paralysis of the heart and lungs, and the adders carry a hemotoxin which destroys blood vessels and clots the blood. The cobra venom is slow in local effect but quite rapid in general effect and therefore a tourniquet should always be applied; the adder venom is rapid in local effect but slow in general effect so that a tourniquet should not be applied. One might ask how a victim who does not know snakes can tell a cobra from an adder "because all snakes look the same." In fact it is not difficult to tell and in any case people living in areas where snakes are about usually know roughly what the difference is. Generally speaking, the adders are thickset and short with broad heads; the cobras are slender and long; they have no visible neck, and the head is inconspicuous.

Is it possible to be immunized against snake bite? A few people have claimed to be immune, but nobody can survive the full bite of a mamba, a krait, a tiger snake, a puff adder, a diamondback rattler, or a score of other snakes unless he receives antivenin. Horses, however, do become immune when they are injected with ever increasing doses during the manufacture of antivenin. Antivenin is made by injecting a horse with a specific snake venom but not enough to kill the horse. When the animal recovers, a large dose is given and then a larger one until finally it is receiving seventy times the lethal dose. Its blood is then tapped and the plasma separated. This clear liquid is the antivenin used for human injections against snake bite.

Which is the more deadly snake family: the cobras or the adders? The hemolytic snakes—vipers such as puff adders, Gaboon vipers, rattlesnakes, and Russell's vipers, which carry blood-coagulating venom—do not kill as quickly as the cobras and therefore a man has more time to get to serum. On the other hand, the bite is a most unpleasant one with very damaging local effects. The toxin may cause immediate intense pain, swelling, and destruction of tissues in the bitten area. There are localized signs of blood vessels rupturing. Then follow the symptoms of shock, such as irregular pulse, perspiration, dizziness, and vomiting. Within a few hours the limb and possibly adjacent areas of the trunk are badly swollen and discolored and within twenty-four hours there may be signs of widespread subcutaneous hemorrhages. Blood may flow from the mouth and nose, and there may be internal bleeding. The victim in fact appears to have been physically battered. Death, if it occurs, may take several hours or even days. There is a secondary danger of coagulated blood in the muscles mortifying, which may be fatal even weeks later. Should the patient survive he might suffer complications for a year and more.

"It is much more pleasant to be bitten by a cobra," an African snake catcher once told me in all seriousness and I can see what he means. The cobra's neurotoxin usually causes sleepiness, weakness, vomiting, and partial paralysis within a fairly short time. The lungs might collapse but the heart action, intensely accelerated, carries on, and convulsions may precede death. If the victim recovers he usually does so rapidly and there are rarely complications. These nerve poisons usually attack the anterior brain, which controls breathing. Should a cobra bite an artery, death would be almost instantaneous—almost as quick as a brain shot.

Antivenins today are usually reliable and cover almost all the world's dangerous snakes. In Africa, for instance, one can buy a polyvalent serum which covers both cobra bites and adder bites with the exception normally of the mamba's and the boomslang's, both of which require a specific antivenin. In cases of bites from snakes which inject massive doses of venom, such as India's king cobra and Africa's Gaboon adder, only an almost immediate injection can save a victim's life. Herpetologist Denis Groves was bitten on his little finger by a Gaboon adder and, although he received polyvalent serum within thirty seconds and was assisted by one of the world's leading authorities on snake bites, his life hung in the balance for twelve hours. His finger was amputated.

In the United States only 3 percent of snake-bite victims who have received antivenin die. Fifteen percent of untreated victims die. Of course, the picture in Africa, Asia, and Australia, where the snakes are far more toxic, would be quite different.

A number of people have been bitten by highly venomous snakes and yet they have not succumbed. This happens quite often, even when the victim has not been treated for the bite. The snake had already expended its venom elsewhere, if it ejected venom at all, or it ejected it into the victim's clothes. Sometimes a snake bites deep but releases no venom. When this happens, the snake's hollow teeth have probably become blocked, or the snake withheld the venom "voluntarily." On the other hand, some people are reported to have died after only a scratch from a serpent's tooth: after all, some cobras can deliver in one quick bite twelve times a fatal dose. In a mere scratch delivered in a split second they might deposit enough venom to kill a man.

Most people assume that the long, whiplike cobras are the fastest-striking snakes, whereas it is the thick, sluggish-by-nature adders which take the honors here. That dreadful customer the puff adder was, for years, thought to be capable of striking backward. In fact, it was turning around, opening its mouth, and lunging forward so quickly that the first movement was undetectable to anything but a high-speed camera.

The world's fifty species of sea snakes are largely confined to the Indo-Pacific area, especially along the shorelines of the Far East and Australia. The yellow-and-black sea snake, *Pelamis platurus,* a handsome reptile with an oarlike tail, is perhaps the most ubiquitous. It is found along the east coast of Africa, where from time to time it kills offshore fishermen who accidentally catch the snakes in their nets. Occasionally these snakes are washed up on beaches or even crawl ashore and traverse land for short distances but, so far as I know, there is no record of one having bitten anybody on land. Most sea snakes prefer the shallows and estuaries but they can be found in the middle of the ocean.

The bites of most sea snakes have a high mortality rate. Although there is no specific antivenin for most species, a polyvalent serum with an elapidine component is customary. The venom is similar in character to the cobra's.

There were men who emerged from the South American jungles with stories of anacondas forty feet long. Some even claimed one hundred feet long, which would have meant they would have a girth equal to a subway train. Then a few years ago the New York Zoological Society offered five thousand dollars for the first thirty-foot specimen. The prize has yet to be claimed and the stories of giant anacondas have died down. The biggest anaconda ever collected, *Eunectes murinus,* was nineteen feet in length, and three feet in girth, and weighed 236 pounds. There was one reliable record of a twenty-five foot specimen but that, for the time being, is the biggest.

A lot of literary license has been spent on the various species of large constricting snakes, and although there may not be any capable of dangling from trees waiting to snatch cattle off the ground (as one author claimed) there are some very large ones.

According to R. L. Ditmars, the record largest were: the reticulated python, *Python reticulatus,* of eastern India and Malaysia, thirty-three feet; the South American anaconda, *Eunectes murinus,* twenty-five feet (but heavier than the reticulated python); the Indian rock python, *Python molurus,* twenty-five feet; and the African python, *Python sebae,* which just exceeded twenty feet. But these records can be regarded as freakishly large. Ditmars says that a reticulated python of twenty-two feet and an anaconda of seventeen feet would be considered large by normal standards. I know that today a nineteen-foot African python would be quite a sensation.

All of these big constricting snakes are hypothetically capable of killing a man and a thirty-three-foot python could even swallow a man. But there is no record of an adult being eaten by a snake. Children? It is difficult to say, but judging by the number of stories one hears around the villages of Africa I would say the African python at least has taken an occasional African child. It seems logical also that the Indian python, which has been found with an adult leopard inside, has probably eaten Indian infants from time to time.

In September 1965 Reuters put out a report about a ten-year-old Singapore schoolgirl who trod on a twenty-foot python, which immediately encircled her. A gardener cut the snake loose and the child recovered in the hospital. Two days later the same news agency put out the following message datelined Rangoon: "An eight-year-old boy was swallowed by a python in Ye village, lower Burma, press reports said today. Villagers killed the fifteen-foot python and took out the boy, but he was dead. His bones had been crushed." The last part of the report is probably pure conjecture, for pythons do not crush their victims' bones: they kill their prey by tightening their coils every time the victim exhales until finally the victim suffocates. The reticu-lated python, *Python reticulatus,* is blamed for at least one death: a fourteen-year-old-boy, who is said to have been swallowed on the island of Talaud in the East Indies. As the python occasionally enters villages in search of poultry and dogs it again seems logical that the species might have taken infants from time to time.

In Africa the average python would generally find it difficult to kill a man, although there is at least one authentic instance on record of its having happened. It involved a Nyasa named Hurly, who worked at the Alpine Mine in the Eastern Trans-vaal. On January 18, 1961, in the hills above the bushveld mine, Hurly saw a python and caught it by the tail. The snake was longer than he thought and it threw two coils around him and began to constrict. Hurly fell to the ground. Man and snake rolled about for some time before the Nyasa managed to uncoil the reptile. He returned to the mine and although shaken showed no signs of damage. The following day he reported to the mine hospital with a bad headache. He was admitted and died the following day. An autopsy showed a ruptured spleen and kidneys. Loveridge records second-hand "a well-behaved woman" being killed in Tanganyika, which is more sad than authentic. Osmond P. Breland records an adult woman killed on the island of Ukerawe on Lake Victoria. In another incident, this time involving a European nurse near Mwanza, also on Lake Victoria, C. J. P. Ionides records that she was thrown down by a python which savagely bit her, causing her serious injury. John Sinclair, the wildlife photographer, was once bitten by a python and he showed me his spiked hand the next day. Judging by the damage one bite did, several bites could be serious.

The best demonstration of a python's capabilities I ever witnessed took place a few years ago when eighteen-year-old Dawie Field, a slight young man of Johannesburg, showed me how to catch a python. He selected a fourteen-foot python weighing 130 pounds which had been captured for the Transvaal Serpentarium. Field provoked the snake into attacking him. At one point he fell into a swimming pool on purpose, and although the snake had three coils around him—one around his neck—Field was never in trouble. When the snake began to constrict Field uncoiled it from the tail and the snake did not have the strength to resist. "The secret," said Field, "is to always keep one hand free. I think if it had pinned both my arms I would have been in trouble." Ditmars says that a ten-foot or longer python or boa *could* be dangerous and one over fifteen feet *would* be able to constrict a man. He says further: "I doubt if ever a twenty-foot python could swallow the average adult human, owing to the breadth of the shoulders, but it is conceivably possible and we take no chances."

Anatomically pythons differ from boa constrictors, *Constrictor constrictor,* by lacking a certain bone in the head, by living in a different environment (the python prefers the bush and the boa the forest) and by laying eggs (boas produce their young viviparously). Both pythons and boas differ from other snakes by having vestigial pelvises, rudimentary hind limbs, and paired lungs.

There are several species of boas ranging from pencil-thin

rosy boas to the giant fourteen-footers of tropical America. Seeing that one giant boa has been recorded as swallowing a ninety-three-pound goat it means that they could swallow a child but there is no instance of man-eating boas that I can find. Boas normally satisfy their appetites with rabbits, birds, and mice.

In England, my father once killed a viper. *Vipera berus,* I suppose it must have been. He told the man next door and demonstrated its size by holding his hands apart. It was about a foot and a half long the day he killed it. When he told the neighbor on the other side he held his hands at least two and a half feet apart and when this fellow's wife came out the snake grew at least another foot. The next day he was demonstrating its size by holding his hand above his head and looking up and down the distance between his hand and the floor. The story got back to me a few days later through a fellow who I never knew had dealings with my father and when I asked him how big the snake was he paced the distance. Today one can only assume that my father wrestled with an anaconda and that he was lucky to have escaped with his life. But that is how it goes in England with the common viper: it's the only vaguely dramatic animal there and people tend to make the most of it. The fact is that *Vipera berus* is relatively harmless and according to Crompton has killed only seven people in half a century (up to 1945)—four of whom were children.

The continent of Europe has another five mildly venomous snakes, (and twelve nonpoisonous) but, as they say, these belong to the "little league" and, far from being a nuisance, are considered to play a vital part in the control of vermin in Europe.

Asia is the home of the king cobra, which Raymond L. Ditmars described as "the star snake of the Indo-Malayan area, *Naia hannah* [now called *Ophiophagus hannah*] . . . not only is it the world's largest poisonous serpent and the most deadly of all reptiles owing to the great amount of a particularly powerful neurotoxin its venom glands secrete, but it is the most dangerous of all living creatures. Combined with the deadliness of its fangs, it is extremely active and commonly inclined to attack. Coupled with insolence, sometimes prompted by curiosity, but more often by danger, is an intelligence that renders it unique."

The largest king cobras—around sixteen feet—are said to come from Thailand, where they have been reported to kill trained elephants which accidentally trample them. They apparently bite the elephants in the soft skin between the toenails. The pachyderms, according to the files of the Academy of Natural Sciences of Philadelphia, die within three hours. It is mainly the terrific volume of venom injected rather than the efficiency of the venom itself that makes the king cobra's bite so lethal. The veteran herpetologist Wesley Howard Dickinson collapsed within three minutes of being bitten by a king cobra while force-feeding one in his home in Santa Ana, California. Within the hour, in spite of advice from leading toxicologists, he was dead. OF MALVERN, PA.

 One day Dan Mannix went to see the snake expert Grace Wiley in California to film her with her newly acquired cobra, which she claimed had a "G" on the back of its hood—"G for Grace." As Mannix looked through the viewfinder he saw the snake spread its hood—something which Grace Wiley had been trying to make it do—and then it struck her. Mannix watched in horror as it chewed on her finger. Miss Wiley gently parted its jaws and returned it to the barn. She wasted several valuable minutes before she showed Mannix where her snake-bite kit was kept. By then nothing could have saved her and she died within ninety minutes.

One really has to see a king cobra to appreciate why such a man as Ditmars would say, "I have always felt a curious respect for it, akin to awe." I once saw a newly captured one— it was twelve feet long and as soon as it was released onto a lawn it reared six feet into the air and spread its magnificent hood. It seemed to tower above me although, in truth, we were eye to eye. It was the most fascinating snake I ever saw and it looked what it is: the most deadly creature on earth. One thing that I did notice was that when it struck it was very slow. I was told this was because it was tired, but most cobras have seemed slow strikers to me. One particular cobra—the African ringhals —which I have often seen striking, lunges forward awkwardly, and if it misses it falls flat on its "face," struggles, regains its upright stance, and tries again.

The so-called spectacled cobra or Indian cobra (*Naja naja*) is the biggest killer in India partly because it is so common. Its bite, provided it gets a good chew, is enough to kill fifteen people. Of the twenty thousand or so people killed in India annually by snakes, possibly as many as 90 percent die from the bite of this snake. Few cobras can be described as slow to provoke, and this particular species is extremely short-tempered. Corbett gives an account of how an Indian was cornered by *Naja naja* in his hut and in trying to ward the snake off was bitten a dozen times. He died within a few minutes.

Ditmars believes that if the natives of India protected their feet and legs when in cobra country the incidence of snake-bite deaths would be halved. This has been proven, he says, on plantations where protective clothing is issued. Another good reason for wearing shoes in snake country, in India or anywhere else in the world, is that one makes far more vibration as one walks along and this is picked up by the snake, which can then move off. There is much less chance of surprising a snake by walking into it. Snakes, without exception, are deaf—they have no ears —and can "hear" only through vibrations coming via the ground.

Second only to the Indian cobra in the numbers of people killed is the widely distributed krait, *Bungarus candidus,* which is nocturnal and has the habit of seeking warmth by lying on footpaths at night while the ground still retains some of the sun's warmth. Again it is bare-legged people who are the worst sufferers. The krait also claims many victims when it tries to sneak into huts at night seeking warmth. Its venom is not as potent as a cobra's, and the snake in many respects can be compared with the coral snakes of the New World. They grow to about four feet, are usually dark and lustrous, and have pale bands. One species has bright-yellow bands, which make it look quite handsome.

There are ten other elapid snakes in India and the Far East but as these are small and totally nonaggressive they are not particularly dangerous. The Hydrophiid sea snakes are most common in Asiatic waters but they too are nonaggressive and human deaths are rare. Ditmars points out that although tourists see them on the sea surface at night when ships anchor off the coast they do not have aggressive intentions but are merely attracted by the ships' lights.

Just as Asia is comparatively poor in its variety of cobras so is it poor in its variety of adders—but again what it lacks in variety it makes up for in individual deadliness. The most infamous of the Asiatic adders is the Russell's viper or tic-polonga (*Vipera russelii*), which Ditmars claims takes a fairly heavy toll of human life. This is the handsome brown snake with black rings which was the villain in Conan Doyle's *The Speckled Band*. Like most adders it is sluggish in habits but exceedingly fast in striking and because of its large fangs it gives quite a volume of venom, which leads some snake experts to say it is as lethal as the Indian cobra. Its toxic effects are slower, though.

Russell's viper lives mainly on rodents and for that reason is attracted to dwellings; hundreds of Indians and Malayans are bitten in their homes each year. Death usually takes several hours.

There is still a great deal of controversy over the species of snake that Cleopatra used to commit suicide, but the odds are it was the asp—or Egyptian cobra, as it sometimes is known (*Naja haje*). The asp would have been a logical choice anyway, for it had deep religious connections in ancient Egypt. Political prisoners were usually given the choice of being tortured to death or dying by the bite of an asp. An odd choice in a way, for the bite of the asp is supposed to be relatively painless, if a little messy.

Paintings of the beautiful Cleopatra lying composedly and

serenely with her hands clasped and her gown smoothed down and with her dead handmaidens draped around are more fanciful than accurate. Although they would all have quickly gone into a coma, death would have been preceded by violent convulsions, copious salivation, and evacuation of the bowels and bladder.

Africa has a dozen different kinds of cobra among its two dozen species of Elapidae and in addition it has three dozen different species of Viperidae. Both families contain some particurarly venomous members and it will be possible here to describe only the extremely dangerous ones. There are more than thirty known killers and a handful of suspected ones. The puff adder is the biggest killer on the African continent, but the black mamba, an elapid which is becoming quite rare these days, probably has the most dangerous bite.

It is an exaggeration to say that the African veld is crawling with snakes. There are certainly plenty about and if you know where to probe for them you can, as a friend of mine did, capture at least half a dozen in a day, including perhaps two or three venomous ones. On the other hand, you can live in parts of Africa and never see a snake. Even in the most infested sections you would be lucky to catch a glimpse of more than half a dozen a year unless you spent your time in the veld. J. A. Hunter, the big-game hunter, in a letter to Denis Lyell said that he saw only about a snake a month on his safaris. On the other hand, John Hillaby, in his journey to Lake Rudolph on the border of Kenya and Ethiopia, passed through a particularly bad area when, it seems, the time of day and the temperature were just right. He saw several carpet vipers—particularly deadly reptiles —on each side of the track. He literally had to watch every step. This type of infestation is most unusual.

I have never seen a cobra in the bush in the twelve years I have lived in Africa and yet they are at least as common as, say, goldfinches in America. The thing about snakes is that they hide away and go into a state of lethargy when the weather is too cool and they hide in rock crevices when the weather is too warm; their temperature tolerance is very narrow indeed.

The asp, for instance, exists throughout most of Africa but it would take you a long time to find anybody who has ever seen one.

The most notorious of Africa's spitting cobras is the black-necked cobra, *Naja nigricollis,* which can eject a fine spray of venom with fair accuracy up to eight feet. The effects are immediately painful if it enters the eyes, and if one has a scratch or a wound of some sort on the face the venom can enter the body with possibly fatal effects. The effect upon the eyes is to cause an inflammation that might last days and can, if the eyes are rubbed, cause permanent blindness. Africans, when one of their numbers is hit by cobra spittle, place the victim on his back and urinate in his eyes. This is fine, especially if water, milk, lemonade, or any other liquid—which will do just as well—is not available.

The cobra's spitting action is so expert that it is uncanny. The snake rears, spreads its hood, opens and shuts its mouth, and there it is. The twin jets of venom are pushed with tremendous force through the tiny holes in the end of tubular fangs and from a distance of six feet it just cannot miss. At eight feet it still is fairly accurate. It can spray quantities of venom three to four times in a few seconds.

A snake which can almost rival the black-necked cobra in spitting is the infamous ringhals, *Hemachatus hemachatus,* a dull, lackluster, black snake—until it rears. Then one sees the high gloss on its underside and the broad white bands across its throat that gave it its name. This cobra is confined to southern Africa and, like the black-necked, has a highly toxic bite. But with all of Africa's cobras, if you carry a snake-bite kit, the chances of complete and rapid recovery, even after a full bite, are very good.

Numbering among the worst of the elapine snakes is the type known as *Dendroaspis,* the mambas. Like the true cobras the mambas also have a neurotoxin but because of the relatively large dose they inject the effects are usually rapidly fatal. The worst is the black mamba, a ten-foot-long, thin, grayish snake with a wide, smiling mouth and two formidably long fangs from which can be milked sixteen drops of venom—two of them are enough to kill a man.

The black mamba, *Dendroaspis polylepis,* is generally regarded as the fastest snake in the world when it comes to traveling over the ground, although it is not a particularly fast striker. A number of people believe it can keep pace with a galloping horse. Bernhard Grzimek once stated that even an Olympic runner could not outrun it. In reality, although it *is* the world's fastest, it can move at only seven miles per hour over a long distance.

The black mamba is a macabre sort of snake: one in the London Zoo—I'm not sure whether it is still alive—was caught by Ionides in 1954 in Central Africa at a spot where it, and possibly its mate, had killed seven people over a period of a few months. Charles Pitman records a "rogue" mamba in Rhodesia which began biting sheep and ended up by killing eleven people, all of whom died within thirty minutes. I recall an incident in 1958 at Umbombo in which a black mamba bit two African girls. Both died in an hour. A few years back a healthy Sprinkbok rugby player was bitten by a mamba, and he took the old-fashioned remedy—a bottle of brandy. He died within six hours. It was the brandy that probably killed him, for he had only been scratched and a little common sense might have saved his life. The most rapid death from a mamba I have heard of is ten minutes, but almost instant death might result from a mamba bite in a large vein.

There are probably more yarns about this snake than about any other snake in Africa. In some ways its reputation and its manner of striking a person high up on the body is reminiscent of the king cobra. One story that is told with regular monotony is about the farmer who was cutting bush with a panga when he grabbed a black mamba instead of a branch. The snake bit his arm. On his next stroke the farmer severed the snake. On his next he severed his arm and so saved his life. It might have happened. Another common story is about a black mamba that fell from the rafters of a hut into the midst of a family of five: they all were bitten and they all died.

There is another moderately common mamba in Africa: the green mamba, *Dendroaspis angusticeps,* which although more arboreal than the black mamba is just as deadly on the few occasions it attacks. Ionides said that on May 23, 1959, near Newala, in the Southern Province of Tanzania, three men, three women and two children died after a green mamba got into their hut. The only survivor, he said, was a baby. Ionides later caught the snake. The fact that this incredible snake catcher, Ionides, caught about 3000 green mambas within a very few miles of Newala shows just how numerous they are and also just how nonaggressive they must be, for not many people are bitten by them.

One of the few dangerous snakes of the Colubrids in Africa is the boomslang. Until 1962 there was no antidote for its bite, and any man who received a full bite was a dead man. It is interesting to see how Ditmars describes them as "mildly poisonous" in his *Snakes of the World.* In fact, most zoologists considered them absolutely harmless until a few years ago and they were kept by children as pets.

The boomslang (Afrikaans for tree snake), *Dispholidus typus,* is a longlish, slender, arboreal, back-fanged snake which, for a snake, has large, quite friendly-looking eyes. It can be olive, bright green, yellow, brown, or shades in between. It is usually very difficult to provoke one into striking and Ditmars' experience in which half a dozen of them emerged from a box lashing and biting is unusual indeed. Weight for weight, according to B. J. Keyter, their venom is the most poisonous snake venom in Africa but, fortunately, because the snake's fangs are set so far back it can hardly get a grip on a person's limb unless it manages to seize the tip of a finger and chew. People who survived the boomslang's bite before 1962 did not get a full bite. The venom is more hemotoxic than neurotoxic and is fairly slow acting. In fact, the patient can be saved with antivenin even after two days in some cases, and the bite normally takes three to four days to kill.

A typical case history is that of herpetologist Bert Mitchley, who was bitten by a pet boomslang in September 1961. Mitchley was bitten on the finger and the snake managed to hold on for only a second or two. He immediately drove to Pretoria General Hospital where some of the world's top toxicologists worked on him. Nothing helped and within three days he was so black and blue with burst blood vessels that he looked like a bad road accident case. He died on the fourth day.

But because the boomslang is so genial and is not really equipped to bite anything bigger than a rodent, very few people are bitten. If they are, antivenin can be flown to them from the only source in the world, the South African Institute for Medical Research, in Johannesburg. The reason for the rarity of the antivenin is the difficulty in getting supplies of boomslangs for "milking." Those that are milked usually die in two or three weeks from lip canker.

Of Africa's vipers the worst by far is the puff adder, *Bitis arietans,* in that it is the most ubiquitous and because of its habits is the snake most likely to be trodden upon. Even a day-old puff adder can deliver a fatal bite. Undoubtedly this snake kills more Africans every year than any other animal of comparable size or larger.

If the Africans wore protection around their feet and legs the number of fatal puff-adder bites would be halved—as would the number of cobra victims in India. The puff adder is known to be responsible for a few thousand deaths a year. A number of its African victims die of fright or shock even in cases where the dose of venom was insufficient to kill them. Panic-stricken African victims have been found trying to drink milk from a cow's teats because they believe it helps neutralize the venom. Others have been found running until they drop because they think that running helps.

These well-marked snakes—they have brown, angular patterns rather like a golf sock—usually grow to two feet. Some have reached almost double that length and they are also quite heavy. I have seen them as long and as thick as a man's arm. They are sluggish when on the move and spend a great deal of time basking in the sun, or keeping warm at night on bare patches of earth—quite often on a footpath. Barefoot children, who set up very little vibration in the ground as they walk, are frequently bitten when they tread on a sleeping puff adder. Once a puff adder delivers a bite it often hangs on and worries its victim as a terrier worries a rat. Its inch-long fangs and tremendous venom sacs enable it to give a heavy injection deep in the wound and such a bite, if untreated, usually kills within a day. A donkey was recorded to have died twelve hours after a bite. In one highly unusual case a person died within two minutes, although this might have been due to shock.

The curved, needlelike fangs of a viper are wonderful things: they are hinged in such a way that when the snake opens its mouth to strike, the teeth fly forward so that they are pointing directly at the prey. In other words, the prey is stabbed before the snake's mouth closes over the punctures.

There are five other dangerous adders in Africa and one of the most sinister is the heavy, exceedingly thick Gaboon adder, *Bitis gabonica,* which has the most beautiful markings, rather like autumn leaves. It packs a tremendous bite, which contains both neurotoxin and hemotoxin in fair quantity. These adders are much bigger than the puff adder and a great deal heavier and have heads shaped like a shovel, the bulging sides of which contain the venom sacs. Fortunately this lethal animal is good-natured and seldom bites people. An acquaintance in Zululand learned this one day when he sat on a log and opened up his sandwich pack. He felt something soft move beneath his feet and saw it was a Gaboon adder. He lifted his feet and the Gaboon slid slowly away. Usually a Gaboon gives a good warning when one gets too near: it emits something between a growl and a hollow hiss which is very audible and unmistakable in meaning.

The saw-scaled or carpet viper of the more northern countries of Africa, *Echis carinatus,* the burrowing adder, *Atractaspis bibronii,* and the berg adder, *Bitis atropos,* are each capable of fatal bites but apart from the first named very few deaths are recorded. The little carpet viper was described by Ionides as unusually aggressive and he mentioned that all 311 he handled tried in the most determined way to bite him.

One more adder has a bad name: the ominous-sounding night adder, *Causus rhombeatus*. There are some terrible stories about this adder. One that is often repeated, but which I do not believe, concerns a farmer who was bitten by one and who sucked the wound to clear out the venom. Because he had a sore throat (or a gum ulcer) the venom caused his throat to swell suddenly so that he choked to death. Apart from this I know of not one case where the night adder has killed and, as far as being a dangerous snake, I would put it on a par with the European adder.

A teamster in the old days was rolling West when a massive rattlesnake lunged at his mule team and missed. But it buried its fangs in the tongue of the wagon—the long boom that runs down in between the mules. The boom began to swell and in order to save his wagon the freighter whipped out his axe and chopped it off.

There are worse stories than this about the rattlesnake, and although J. Frank Dobie told this one with his tongue in his cheek, the fellow who told it to *him* might really have believed it. In Africa where there are thirty or more varieties of dangerous snakes, people don't talk about them much, but in the United States, where there are only eighteen different species, people seem to talk about them frequently.

The one type, the rattler, dominates the herpetological scene, which is not surprising because there are fifteen species of them and they are responsible for nearly all the 1800 poisonous snake bites recorded annually in the United States. Of these, not many more than 100 prove fatal, and in some years the rattlesnake is blamed for every single death.

It might be best to discuss the other three kinds of dangerous snakes first. The coral snake, *Micrurus fulvius,* is the most toxic of all snakes in North America, including the rattlers, but because it is relatively rare and in any case confined to the South, very few people are bitten by it. Because its fangs are not suited for biting big animals like people, it does not always get in a good bite and so a lot of people survive by default.

One other reason for its lack of human victims is that it is a very brightly colored snake with gaudy bands around the full length of its body. This coloration serves as good a warning system as the rattle of the rattlers. The coral snake is really a degenerate cobra in a sense. Its venom, as is to be expected, is neurotoxic.

The copperhead, *Agkistrodon contortrix,* is closely related to the rattlers and is rather a pretty snake with rich chestnut markings on a light background. Few Americans recognize its beauty nor do they appreciate its docility. In fact, it is so docile that some people have found them endearing pets. The two- to three-foot serpent can barely be counted as dangerous as its bites, untreated, are only .3 percent fatal. In one year, out of 308 bitten, none died—and yet the majority did not receive specific antivenin.

The water moccasin, *Agkistrodon piscivorus,* is short-fanged for a viper-type snake, but a numer of bites are on record in the South, to which area it is confined. The northernmost point in its range is the Mississippi Valley. It is a dull-colored member of the rattler family which is pugnacious, quite venomous, but fortunately rarely come across. This snake is also frequently called the cottonmouth, and is much sturdier than its closest relative, the copperhead.

Of the true rattlesnakes (Crotalidae), it is the prairie rattler (Crotalus confluentus) that reaches the farthest north, extending along throughout the Northwest quite a distance into Canada.

The two most deadly rattlers are the eastern diamondback (*Crotalus adamanteus*) which can grow to eight feet and which is rated among the top twenty most dangerous snakes, and the western diamondback (*Crotalus atrox*) which is shorter and stockier than the former. The western diamondback probably kills more people, for it is more numerous and is found in more populous areas. One's chances of recovering from its bites, however, if the right sort of help is at hand, are very much greater than with the two great diamondbacks.

All rattlesnakes belong to the family Crotalidae, which comprises the pit vipers. Pit vipers are similar to the typical vipers except for one or two differences, the most noticeable being the pair of "pits" or depressions in the front of the head which look like nostrils. The rattlers carry a fairly typical adder venom.

The United States has a great many snakes both in variety and in density but at least one hundred of them are absolutely harmless.

Ditmars in his book *Snakes of the World* recalls a story told to him by Douglas March of the serum station at Tela, Honduras, about the fer-de-lance, the most widely distributed venomous snake of tropical America. A woman was selling sweets on a station platform when she was called away from her stall to attend to her husband, who had been bitten on the ankle by a fer-de-lance. The woman bathed the running puncture marks but that evening her husband died. The next morning the woman died—after displaying typical haemotoxin symptoms. It was found that she had been grating coconut at her sweets stand before she had been called away and had numerous tiny scratches on her fingers. These had absorbed some of the venom around her husband's wound.

The lance headed tropical snake, the fer-de-lance (lance head), belongs to the genus *Bothrops*, which takes over where the North American rattlers leave off around southern Mexico. It has many names, depending upon where one is; barba amarilla (yellow beard) and jararaca (arrow) and fer-de-lance are the most used. The slender, dull-colored snake is the main scourge of Central America's plantations and if its bite is untreated its victim can die rapidly—in fact, of all three dozen species of *Bothrops*, this snake, *Bothrops atrox*, is by far the most toxic.

Tropical America boasts one other really dangerous snake and that is the bushmaster, *Lachesis muta*, the world's biggest viper. An extremely big bushmaster can be twelve feet long. It possesses a particularly powerful neurotoxin, which paralyzes the neck muscles so that the victim has no control over his head. South American Indians swear that somehow the snake contrives to break a man's neck for they can find no other reason for the way the victim's head lolls around like the head of a badly stuffed doll.

This viper, which ranges from Central America right down

through Brazil, is capable of delivering a particularly heavy dose of venom and because of its rather long fangs the venom is pumped in deep. Although there is a reliable antivenin for its bite, the mortality rate is high.

Central America is also troubled by an especially venomous relative of the northern water moccasin, the tropical moccasin, *Agkistrodon bilineatus,* and by several species of coral snakes, *Micrurus,* which, although noticeably amiable and very easy to see, can occasionally cause deaths.

America has its diamondback rattler, Africa has its black mamba, Asia has the king cobra, and Australia has what was once described as "the darkest of them all"—the tiger snake, *Notechis scutatus.* In an unsolicited testimonial Ditmars wrote the following: "This is the most savage and dangerous of the Australian reptiles. Investigations of its venom indicate its being of such extremely high toxicity as to be unmatched by that of any other known serpent. Applied against this, however, are smaller poison glands than a number of the world's better known deadly types. It appears to produce more fatalities in Australia than the combined bites of all the other poisonous serpents in that country."

In a country where there are no fewer than eighty species of elapid snakes (there is not a single species of viper), this is praise indeed. Yet the snake's star billing is not entirely deserved. Australia's infamous death adder, *Acanthophis antarcticus*—in spite of its name it is *not* a viper—kills half the people it bites, whereas the tiger snake kills only 40 percent. The point is, I suppose, that the tiger snake bites more—four times more than the death adder—and kills just over three times more.

The tiger snake earns its name through having "tiger stripes" along its thickish, two- to three-foot-long body. It is a fast and savage snake which is quick to strike and does so with such speed that it shoots itself forward for some distance. Its venom is neurotoxic, as is to be expected, and it can kill a man within two or three hours. There is an effective antidote.

The death adder is unique in the world and is actually a cobra-type snake masquerading as a viper complete with thickset body, short length (two feet) and broad head. It also has the long fangs of an adder and the large venom sacs, which can eject a prodigious amount of venom. This makes it the most toxic snake in Australia, a country which is unique in having more toxic snakes than harmless ones. The death adder is not aggressive and, again rather like the real adders, is sluggish in habits and spends a great deal of time lying on hard ground, especially well-beaten footpaths. When stepped upon it immediately strikes deep and hard. It is a dull-colored snake of the dry areas and is found throughout Australia with the exception of south Victoria.

The two-tone brown snake, *Demansia textilis,* which is 8 percent fatal even though it gives only a small bite and ejects little venom, kills about as many people as the tiger snake. But then it bites 50 percent more people. The black snake, *Pseudechis porphyriacus,* a beautifully colored blue-black snake with a satin quality to its skin, kills only .8 percent of its victims although it bites more people than all the other snakes put together.

The taipan, *Oxyuranus scutellatus,* the notorious and very deadly snake of the northeastern areas of Australia, is fortunately extremely rare and very few bites have been recorded. The majority of its bites have proved fatal when untreated.

On top of all this, Australia has several sea snakes off its coasts, all of which are dangerous.

One would be forgiven for believing that on this massive island people are being struck down by snakes as fast as the birth rate will allow, but in fact this is not the case. The dangerous snakes of Australia bite about a dozen people a year. Of these one, or perhaps two, die.

Looking at the world's array of highly poisonous snakes one cannot help wondering why nature equipped them with such powerful venom when all they need it for is to paralyze or kill small animals such as mice, rats, and frogs. Even if they could eat men their venom is still far in excess of what they would need. Snake venom is certainly not for defense purposes; it is far too slow-acting for that. Even the fastest-killing snake is quite unable to fell an attacking animal quickly enough to save its own life. Its venom, powerful though it is, is still almost useless against owls and eagles, which kill snakes almost with impunity. The snake depends upon its hiss or rattle to warn animals from approaching but this often invites attack. Dogs, cats, and several species of wild animal will go for a rampant snake, killing it long before the effects of the snake's bites kill them. What logical reason can there be for the evolution of the king cobra's venom or the puff adder's? Snake venom, a mysterious cocktail of protein and enzymes, is one of the fascinating zoological puzzles which will exercise zoological minds for years to come.

Do people and some animals have an inborn fear of snakes? The answer is apparently no. Experiments have shown that children or young animals which have never been taught about snakes are neither repulsed by them nor afraid of picking them up. It is a pity in a way that we are so afraid of snakes, for not only are nine out of ten species harmless, but all snakes, "bad" or "good," are vital in controlling the world's rodent population, which has, in man's history, caused far more death and misery than snakes. It is also worth bearing in mind that there are no snake vegetarians and not a single snake is deemed to be an agricultural pest.

VII
MAN-KILLERS
OF THE WATER

The Great White Shark

Fish and Sea Mammals

"You gentlemen of England
Who live at home at ease,
How little do you think
On the dangers of the seas."
Martin Parker

To some people the sea is their life and the fishes in it are their means of making a livelihood. To others the sea is a medium—a very fickle one—on which they travel in ships. To others it conjures up a yacht race, a leaping blue marlin, a seascape, a holiday, a sand castle, or prawns for dinner. To the geographer it is 70.8 percent of the world's surface and deeper than Everest is high; to the biologist it is where all life began; while to others it is a potential source of power, of water to irrigate deserts, and of food for a protein-starved world. The possible uses of the sea are so tantalizing that it has now become imperative for us to explore it on a deeper and wider basis. At the moment we know the features on the back of the moon more precisely than we know the features of the ocean beds and our present groping method of exploring what is under the sea is rather like trying to explore the earth's surface from the sky through an impenetrable layer of cloud.

But the growing body of men who don diving masks and breathing apparatus to grope beneath the sea are facing real danger just as surely as the primitive hunters must have done when they timorously entered the hostile jungle before its carnivores had learned to fear man.

The shark has no respect for man. It will often make a meal of him just as readily as it would have taken a fish. The nervous moray eel will rip at a man's limb just as bad-temperedly as it would snap at any other creature that disturbed it. Barracudas will nose up to divers to see what they are doing and hosts of little fish will sometimes crowd around. In fact, although in some well-hunted areas a few species are apparently learning to avoid spear fishermen, the overwhelming majority of the sea's creatures have not yet learned, as the land animals have, that it might be best to avoid man. Of course, they have no valid reason right now to fear us, for under the sea we are defenseless against the big sharks and creatures such as the killer whale could swallow a man whole. We have never really had to face anything quite like this on land.

Man is on the threshold of discovering the undersea and getting to know its wildlife, but before we can be at ease down there, there will be a price to pay, a price that can be paid only in men's lives.

Considering the diabolical array of defensive and offensive weapons found in sea creatures a surprisingly small percentage of them have a well-developed venom apparatus. Just as on land, so in the sea those creatures which possess powerful venom use it against man only when provoked. Not one poisonous animal in the sea can be described as aggressive toward man. Among the most notorious of the stinging fish are the scorpion fish (Scorpaenidae), which are common along the coasts of warmer countries and in some cases are almost as deadly as snakes. Some of them are so cryptically colored and so stationary in their habits that they are virtually invisible as they lie in rock pools and this is why so many bathers step on them. Perhaps the worst of them all is the stonefish or devilfish (*Synanceschthys verrucosa*), a drab-colored, one-foot-long fish that has venom-tipped spines so sharp and strong that they can easily pierce a beach sandal. A surprising number of skin divers swear these fish can kill within a matter of ten minutes, but as far as I can ascertain, very few victims die and when death does result it is only after many hours of pain. An acquaintance was pierced through the side of his foot off the Mozambique coast and, remembering that he had been told the sting could kill in ten minutes, he waited with some impatience and a great deal of pain for death to put him out of his misery. A companion withdrew the long sting and sucked the wound thoroughly. He also applied a ligature above the knee. After an hour or two the pain subsided and the victim ate a hearty meal. The only aftereffect was a large boil below where the ligature had been.

I should imagine that the very few deaths the stonefish appears to cause would be due to shock. This does not mean to say that the stonefish's venom is not capable of producing death in itself: toxicologists say it can. An antidote has been developed by the Australian Commonwealth Serum Laboratories by extracting the venom of the cone shell.

The venom carried by the shell family *Conus* is similar to the stonefish's venom, which has been described as an hallucinogen that brings about "feelings of deep depression followed by coma" (Edmund H. Burke).

Richard Lurie says that five species of cone shells have caused fatalities in the Pacific and gives their names as textile, court, geographer, marble, and tulip cone. The shells inflict the sting with a long, slender, barbed "tongue" and the symptoms, says Lurie, are "failing eyesight, coma, and death in about five hours." Although species of cone shells are found along the eastern coast of Africa and skin divers go in fear of them, there have been only two or three deaths in recent years.

The only other shell that is supposed to constitute a danger to man is the plankton-eating giant clam or tridacna. It can grow to four feet long and weigh up to a quarter of a ton. Divers treat them warily in Australia, the East Indies, and off the east coast of Africa. Burke states that if a person gets his foot caught in the steel-trap jaws of a giant clam the only escape is to knife through the adductor muscles. Hypothetically a clam could kill a man by holding him under water until he drowns, but I have never heard of such a thing; and it might be significant that Lurie, in his discussion of clams, does not mention this aspect of it in his area.

Another order of fish that must be classed as dangerously venomous is Batoidei, the rays. There are six families of ray that are venomous but among the worst is the stingray family,

Dasyatidae. The rays are flat fish (the skate is one of them) which lead rather sedentary lives half buried in sand in the shallows along most coastlines in the world. According to Burke, there are two thousand cases a year of bathers stepping on them and being stung by the sharp, serrated spine which is found in most of them near the base of the tail. The spine in some stinging types of ray is connected to a poison sac and is similar in some ways to the stonefish's spine in that it is as sharp as a sliver of glass and can penetrate the sole of a shoe. Should it break off inside a wound the results can be serious unless the stinger is removed immediately. In either event, if one is stung by a ray it is best to clean the wound quickly, apply a tourniquet, and seek medical assistance. J. L. B. Smith says that on the African coast even small wounds can be serious. Some result in permanent injury to the affected limb. There have apparently been no deaths from stingrays in temperate waters but there have been one or two in the tropics.

The best way to avoid these fish, which are really modified sharks, is to enter the water with a shuffling walk; once one is swimming there is no danger from the fish.

The electric ray, of which there are twenty species, is found more or less through the oceans and has a bad name which it hardly deserves. It is capable of giving a bather a slight shock but cannot be considered deadly.

The massive manta ray, that huge, flapping devil fish that appears to fly through the water like a gigantic bat, is as harmless as it is frightful. Often called the giant ray, they are really flattened sharks, just like the smaller members of the order, but they are entirely nonaggressive and live on plankton. If ever a manta ray has killed a man the only way it could possibly have happened is for the ray to have taken a flying leap out of the water, which they occasionally do, and land in a boat. As they can measure twenty feet across the "wings" and weigh up to a ton and a half, the damage would be impressive.

Another much-maligned creature of the sea is the mean-looking moray eel. Julius Caesar is supposed to have kept morays in huge pools and fed them live slaves for the amusement of his guests. It would have been a simple matter for a group of moray eels to have been trained to eat human flesh and then starved to the point where they would have lunged at anybody thrown to them.

In nature, however, the moray is not half so bad as painted. Most morays (family Muraenidae) grow to about four or five feet, although some giants of ten feet have been recorded, and weigh an average of, say, forty-five pounds. If one gets too near their holes they are apt to lunge out and take a single but really vicious bite that can scar a man for life. Although the victim normally gets away with a badly torn arm or leg, some victims might possibly be drowned by a large moray because of its habit of hanging on to its victim like a bulldog until beaten off. Beating off a stubby, eight-foot eel would not be easy. There is one other way in which the moray is said to be capable of man-killing: by using its venom. There is little doubt that at least five of the eighty known species of Muraenidae possess fairly efficient venom but there is some controversy over its toxicity. One authority claims it is 10 percent fatal when humans are bitten but I have been unable to find a single recorded instance of a person being killed by moray venom.

The moray is a little bit like the conger eel—big, toothy, difficult to kill, and extremely powerful and most intimidating when confronted nose to nose. It is difficult to believe that either is a man-killer under normal conditions and the evidence against them is very thin indeed.

The legendary, tantacled, evil-eyed, ship-crushing kraken that has haunted sailors' dreams for centuries is, in reality, the giant squid *Architeuthis*. This massive, swift creature of the gloomy, near-freezing ultra-deeps can span fifty feet, has teeth capable of pulverizing a man, and is the biggest invertebrate in existence—but it is not a man-killer. There is not a single proven case against it. This might well be because man, at the moment, has little business to do at the depth at which the squid lives and the squid only on rare occasions swims near the surface.

Victor Hugo probably damaged the reputation of these

strange, usually inoffensive animals in his book *Toilers of the Sea,* in which he describes the man-eating octopus: "You enter the beast. The hydra incorporates itself with the man; the man is amalgamated with the hydra. You become one. The tiger can only devour you; the devil fish inhales you. He draws you to him; and, bound and helpless, you feel yourself emptied into this frightful sac, which is a monster. To be eaten alive is more than terrible, but to be drunk alive is inexpressible." It is a magnificent piece of hokum. He strained the point rather by describing how the double rows of suckers feel—"like so many mouths devouring you at the same time." But if Hugo's prose sends agreeable shudders down schoolboy spines then the truth about the octopus will make them snap their fingers with disappointment.

Jacques Cousteau and his fellow divers found the octopus pathetically frightened of man "and most bashful." He says "they seem as averse to our flesh as we are to theirs." His colleague Didi tried to get one to slap a tentacle around his arm and after a struggle succeeded in getting it to grip him—but the grip was easily broken and the suckers left momentary marks. Burke found that the octopus was "rarely if ever a hazard to divers."

If the octopus has ever killed a man then it did so by poisoning him, for it has in its saliva a fairly efficient poison —a neurotoxin. It can inject this venom by puncturing a man's skin with its parrotlike bill. Burke claims that there have been fatal bites recorded, and as some Pacific octopuses (*Octopus appollyon*) can grow to span twenty-eight feet and must pack quite a dose of venom it would be remarkable if there have not been one or two human deaths. Personally I was unable to trace a single instance.

One of the most unpopular, but at the same time fascinating and quite pretty, venomous creatures of the sea is the bluebottle or Portuguese man-of-war (*Physalia pelagica*). It looks rather like a floating jellyfish but is, in fact, a member of a related group known as *Siphonophora.* The fascinating thing about this creature is that it comprises several animals—a sort of colony of quite different creatures—all of which perform different functions, such as swimming, reproducing, feeding, floating, and stinging. The whole comprises a transparent bluish, floating bag of gas with a high ridge along its top which acts as a sail. From the float dangle threadlike tentacles of which a few are armed with several tiny organisms. When touched they cause the most intense pain. Craig Phillips and the usually phlegmatic Cousteau as well as many other marine biologists describe the animal as dangerous; indeed some authorities claim it is the most dangerous of all these types of creatures. Although the nature of its venom is not clearly understood it can cause severe symptoms which, if untreated, could be fatal if only through shock. I should imagine fatal results must be extremely rare, for the Portuguese man-of-war is very common indeed on the African east coast; and I have often seen them on crowded beaches, where they are frequently trodden upon, yet I have heard of no fatalities. The floats burst with a loud "pock!" and almost immediately one feels a dreadful burning sensation which can continue for hours. It is probably far more dangerous to swim into one and have its tentacles drag over you—tentacles which, incidentally, can grow to twenty feet long in exceptional cases.

A far more dangerous creature than the man-of-war is the sea wasp, *Chironex fleckeri*, which carries a sting capable of killing someone within minutes. One man, stung while bathing off the shore of Queensland, is said to have died within thirty seconds. I remember reading of an incident when I was in Australia in 1962: a priest ran screaming from the sea and horrified bathers saw a sea wasp draped across his back. The creature is similar to a jelly fish. It was peeled off but the priest died an hour or so later. Burke claims that its toxin is so potent that it normally kills within ten minutes. Dr. Robert Endean, zoologist at Queensland University, who isolated the venom for the first time in 1966, said that it was "one of the deadliest toxins known to man" and it seemed useless to develop an antidote, for the animal killed too fast. He thought the best line to take was to develop a vaccine—but this might not be possible.

The sea wasp is said to have caused "several deaths" off the northeast coast of Australia north of Bowen, and the animal is particularly active between December and March. The Cairns area records about ten attacks a year. Fortunately the majority of victims do not receive bad stings, and after such unpleasant symptoms as excruciating backache, chest pains, difficulty in breathing, and perhaps vomiting, they fully recover. The venom carried in the animal's microscopic stinging barbs is similar in character to the neurotoxin carried by cobras.

There does not appear to be any record of the world's largest jellyfish, *Cyanea*—which is eight feet across and has tentacles that spread over seventy-five feet around it—having killed a man, although it has thousands of stinging organisms.

Among the vertebrates both the sawfish and the swordfish are charged with man-killing. The sawfish, which is a true ray, is abundant in the Indo-Pacific area and is also found in the Mediterranean and Atlantic. It can grow to twenty feet and carry a massive saw up to six feet long and twelve inches across at the base. Such a saw would be capable of killing a man with a single blow. J. L. B. Smith claims there is an incident on record of one actually cutting a man in half, but normally this fish is timid and similar in behavior to the smaller rays.

The swordfish has also been indicted but the evidence against it is, like that against the sawfish, most unsatisfactory. Of the family Xiphiidae, which includes some good game fish, the swordfish can grow to fifteen feet and, at that length, would have a sword some three feet long. The sword is a tremendously hard, wickedly pointed prolongation of the fish's upper jaw and it is used for spearing prey. The fish can charge at such a speed that the sword has been known to pierce clean through the bow of a boat. In Britain there is a museum exhibit of a twenty-two-inch swordfish's sword, which has passed thirteen and a half inches through a ship's plank. Breland claims that in 1830 a man was fatally speared off England, and there might have been other freak cases.

One of the most controversial of fishes is the barracuda, *Sphyraena barracuda,* which is also called the picuda and the becuna. This thin fish is long for its size—up to ten feet—and resembles a pike right down to the razor-sharp rows of teeth. In some parts of the world skin divers and others fear it more than the shark, but this reputation may also be worse than it deserves.

I once asked a free diver who operated off the dangerous coast of Portuguese East Africa what he considered the most dangerous fish in the area. Unhesitantly he replied: "Sharks, then 'cuda." I asked him if he knew any divers who had been attacked by barracudas. He said: "Not personally. But you only have to look at their teeth. . . . I don't mind the odd one who comes up to see what's going on. It's when a pack approaches, that's when I get out." Coincidentally, not a week later, in January 1967, a Portuguese soldier, Private Alberto Carvalho, was swimming in clear water off Lourenço Marques' most popular beach when he was grabbed by a barracuda which stripped a large piece of flesh from his leg. The soldier swam to a floating tree and clambered into its branches with the fish still leaping at him. The fact that the 'cuda came back for more is most unlike these fish, which normally flash in and out and are gone.

Burke states that "one can never be sure with a barracuda," which just about sums it up. He says that since 1884 there have been fourteen attacks in the United States definitely attributed to this fish. Undoubtedly a lot of supposed "shark attacks" are the work of barracuda. There seems to be no record of a barracuda having killed a man, although it is hypothetically possible for a pack of them to do so or for even one of them to inflict a mortal wound. Curiously Cousteau claims, "Barracuda are no danger to divers," and adds, "I know of no reliable evidence of a barracuda attacking a diver."

Before examining the sea mammals, we must turn to a few fresh-water fishes that rate as man-killers.

Eduardo Barros Prado in his book *The Lure of the Amazon* has a chapter titled "The Girl with Red Hair." The chapter tells the story of Leena, a pretty redhead who dived from her uncle's yacht into the mouth of the Canuma in Brazil for a swim. Suddenly, a short distance from the yacht she stopped swimming and disappeared below the surface. "A moment later," writes Prado, "she reappeared, her arms thrashing the water and with terror on her face. She screamed and the water around her became alive with movement." Nolan, the captain of the yacht, immediately moved over to the girl but by the time he got there she had disappeared. "A few ripples broke the surface and a faint crimson tinted the water. Soon this dissolved and disappeared, leaving the pool placid and empty."

Prado and the others made a search and later found the girl's skeleton in a torn scarlet costume; left untouched was her flowing red hair.

Prado explains what happened: "Leena had been attacked by deadly flesh-eating piranhas, the small fish which attack any living creature, particularly humans. They hunt in shoals and so great is their voracity they can reduce a living person to a skeleton in a few minutes."

I remember years ago reading in some magazine about a crew member of a Brazilian riverboat being punished in a very corporal sort of way by being jammed into a life belt, having his hands tied behind him, a small nick made in his leg—and then being tossed over the stern to bob up and down in the wake. As I recall the story he screamed briefly and then, suddenly top-heavy, turned turtle in the water, exposing his lower body which was just a skeleton. Lieutenant Colonel P. H. Fawcett tells a similar story of how, during some civil strife in Brazil, the locals tied their prisoners to stakes, waist deep in the river, and then giving the traditional nick in the flesh left them to the piranhas. In his hair-raising book *Exploration Fawcett,* the colonel also tells about a red-trousered Brazilian soldier who fell out of a canoe into the Corumba. When helpers went to haul him in they found that the fingers gripping the gunwale were a dead man's fingers: the man's body below the waist had been eaten clean away, red trousers and all.

There are so many of these stories and, although some suffer from the heavy hand of exaggeration, the gist of them is usually true. Ever since Theodore Roosevelt wrote of his experiences in South America and mentioned the cannibal fish (his name for the piranha) the fish has preyed upon the public mind. Two of the President's party suffered from a piranha attack: one lost a toe and one lost a few ounces of leg. In a paragraph the President summed up the voracity of the fish: "One of the most extraordinary things we saw was this. On one occasion one of us shot a crocodile. It rushed back into the water. The fish attacked it at once and they drove that crocodile out of the water back to the men on the bank. It was less afraid of the men than of the fish."

There is little doubt that in some localities, especially in the vicinity of villages and towns where animals are slaughtered, the great shoals of piranha fish have got used to rushing at anything as soon as it falls in the river, and one must assume that goes for humans too. But apart from such places it is said to be safe to swim in rivers where piranha fish are known to be. In fact, according to Grant White, they are so scarce in the Amazon River today that he spent months searching for some to film. When he eventually found some he was not disappointed at their acting abilities. He shot an iguana, cut its tail off, and cast the animal over the side. "Within seconds," says Grant White, "there was a tremendous commotion and the river began to boil and bubble in an extraordinary manner, just as if it were a seething cauldron. We then saw it was caused by hundreds of piranhas which were all making a concerted and ferocious attack on the corpse of the lizard. In a trace it was completely devoured and all was quiet again."

White describes the fish as between nine inches and one foot long with silver scales splashed with scarlet and weighing about a pound each. There are actually several species of piranhas but the ones that rate as man-eaters are mostly about a foot long, flattish and deep with a heavy and powerful lower jaw which juts out in a most determined way. Its mouth is primed with tiny needle-sharp teeth. The really notorious ones are of the genus *Serrasalmus,* and the various species differ in many ways, particularly in size. It is a rare occurrence indeed today to hear of a man being killed by these fish, mainly because of a tighter control over river pollution.

I think that Burke, when he says that "the piranha will unhesitatingly attack anything in the water," is generalizing. So are the many writers who subscribe to the belief that some people who fall in the water are eaten so quickly that by the time they surface they have been torn to pieces and within seconds their skeletons are picked clean. Blood certainly excites the fish but even Roosevelt allowed that it took minutes for them to strip a shot duck.

Roosevelt also brought to the American public's notice the man-eating habits of the giant South American fresh-water catfish *Pangasianodon gigans.* This fish has often been accused of trying to nudge boats over and is most certainly a monkey-eater, snatching them from branches as they hang down over the water to drink. Roosevelt's party actually found a catfish with a monkey inside it. This led one of his Brazilian friends to tell him "that in the Amazon there is a gigantic catfish nine feet long. The natives are more afraid of it than of the crocodile because the crocodile can be seen but the catfish is never seen until it is too late." The President was told that villagers built stockades on the edge of the river so that their womenfolk are protected from the fish when they collect water.

Catfish, even small ones, are pugnacious creatures and are carnivorous. Although I have yet to trace an authenticated incident of this South American monster's eating a person— children are said to be its preference—I would still prefer to be out of the water when one is around.

I have seen giant catfish (called barbels) in South Africa but to my knowledge nobody is nervous about them. There is, however, a giant catfish found in the rivers and estuaries of central and eastern Europe and western Asia which is reputed to reach a length of ten feet. It attacks almost anything that it can swallow and has been known to attack children bathing in rivers.

Although the chances of a catfish fatally injuring a child are remote, there is always a danger from the poisonous spines which most catfish have. Here is yet another imperfectly understood venom, which, at least hypothetically, could kill a small child or a weak adult.

One would expect at least a little shock from the South American electric eel, *Electrophorus electricus,* with such an impressive scientific name. In truth, the shock can be as big as five hundred volts at two amperes (a thousand watts) which has been enough to kill a horse. But there is no record of one having killed a man. The electric eel is not a true eel but is more allied to the catfish and shares a family (*Gymnotidae*) with the knifefish. Four-fifths of its body is tail, and it is in the tail that the "batteries" are housed. The fish can grow to eight feet and be as thick as a man's thigh—but it very rarely does. Few exceed four feet. Hypothetically it could kill a man by stunning him and thus causing him to drown.

In 1891 a whaler closed in on a massive sperm whale which suddenly turned on the boat and smashed it. The crew were flung into the foaming sea, and later all but one of them were rescued. A few hours afterward a sperm whale was harpooned from another open boat and was dragged alongside the whale ship. Next morning when the flencers were opening the beast they reportedly saw something move in its stomach. This would not have been unusual, for the sperm whale will occasionally swallow giant squids which can take a long time to die. On opening the stomach wall the whalers found the missing man, or so the story goes. He was bleached white by the whale's gastric juices wherever his clothes had not protected him and, says the fellow who started all this, he remained very white for the rest of his days.

I personally doubt the story, but I rather like it. But it is hypothetically possible for a sperm whale to swallow a man, and judging by the hair-raising methods used by whale catchers in the old days to catch the sperm whale (or cachalot, as it is called), it would be surprising if none were swallowed even by accident. There are plenty of incidents on record where whale hunters were killed when hand-harpooned whales smashed small open boats into matchwood.

A yachtsman who had taken part in the annual race around the Cape from Cape Town to Port Elizabeth told me once how their two-master was becalmed and the skipper told the crew they could have a swim. They swam around the yacht until the skipper saw a breeze discoloring the sea's surface and he called his crew to jump to it. They scrambled aboard just as the sails bellied out and the yacht leapt forward like a live animal. Suddenly everybody was thrown forward as the yacht shuddered to a halt with a mighty jump. A second later the sails shook out again and the vessel gathered speed once more. The skipper stripped the masts, and two of the crew went over the side to find out what they had hit and what the damage was. They found not a mark. The yacht finished second in the race that year and then returned to her home port in Cape Town, where the skipper thought it best to drydock her and have a good look at the hull. It was then they discovered a two-foot-wide mouthful of wood was missing from the wooden rudder. They also found teeth embedded around the missing portion. They were identified as the teeth of a killer whale.

The title of the biggest confirmed man-eater on earth must surely go to this thirty-foot-long member of the dolphin family, the black-and-white grampus or killer whale (*Orcinus orca*). It could, if it ever had the opportunity, eat a couple of dozen men in one session. Eschricht records one which was found with thirteen porpoises and fourteen seals in its stomach—it had choked on the fifteenth seal. A killer whale was once shot off the West Coast of the United States and found to have eighteen fur seals in its stomach, each one heavier and bulkier than a man. There is no animal that swims in the sea or accidentally falls into the sea that will not make a meal for a grampus. They will attack the largest whales, and are said to tear out their tongues. Burke writes: "Nothing in the water or on the land can compare with the killer whale for sheer ferocity and ruthlessness. They are absolutely fearless and hunt in packs." They are often referred to as the "wolves of the sea" and the Eskimos live in fear of them and will not even point a gun at them. They believe that they can turn themselves into wolves just as wolves can turn themselves into killer whales.

The Eskimos have good cause to fear them for the whales sometimes mistake their kayaks for one of their favorite prey, the small beluga whales. They toss the kayaks high into the air, says Sally Carrighar, in much the same way they toss the belugas up before catching and eating them. There are numerous stories of killer whales attacking and bumping even quite large boats and no one will ever know how many shipwreck survivors during the war were upset from life rafts and eaten, especially in the North Atlantic.

They are no less a menace in the southern oceans, and the U. S. Navy's *Antarctic Sailing Direction* says bluntly, "Will attack human beings at every opportunity." Herbert F. Ponting, the photographer on Scott's last expedition to the Antarctic, saw eight of them swim under the pack ice and begin to bump it with their backs in the hope of breaking it so that he would fall in. One even rose from the sea and looked along the top of the ice at him as it blew off a spray of fishy-smelling vapor.

Ponting described their heads as lizardlike. They have in fact a fairly massive head, very black on top and very white underneath and armed with a pair of formidable jaws bearing forty needle-sharp teeth. They are unmistakable when swimming or basking along the surface of the sea because of their high dorsal fins, which can be five feet high and resemble the tailpiece of an airliner.

A potential man-killer down in the Antarctic is the sea leopard or leopard seal, *Hydrurga leptonyx*. Walter Sullivan says its reputation rivals that of the killer whale. It earns its name from its spotted fur and might have earned its reputation from having chased small boats now and again but in fact there is not a scrap of other evidence to suggest it is a man-killer. It grows to about twelve feet long.

Sea lions have savaged bathers from time to time off the coast of South Africa and it is difficult to see their motive. None of the attacks has been serious.

According to Alwin Pedersen, walruses have killed men. He says that a "wounded walrus is the most dangerous animal in the Arctic. Instead of escaping it attacks furiously." On land this animal, huge as it is, should not be difficult to escape from but in the sea a man would stand little chance. Pedersen mentions how eight men were lost in a sea battle with a number of walruses and says that one walrus in trouble might well be aided by others. Many Eskimos have probably lost their lives to walruses—especially before the advent of the gun.

The sea elephant, *Mirounga angustirostris,* can be dangerous, especially old bulls. In March 1967 one ripped open the bow of a small boat off Napier, New Zealand, sinking it. The sea elephant swam around the five crewmen as they swam ashore but did not attack.

John Pitts witnessed the pathetic attempts of an old bull, which charged him on the ice shelf on Antarctica. The bull was shot for husky meat. "It was as pugnacious as hell," I was told, "but so laborious were its movements that one could easily walk out of harm's way. In the sea, I couldn't help thinking, it must be dangerous to men in a dinghy. The animal weighed a ton."

The Shark

"But when the tide rises and sharks are around,
His voice has a timid and tremulous sound."
                                    Lewis Carroll

During the Second World War, a Navy flyer, Lieutenant Reading, and his radio operator, Almond, ditched in the central Pacific. Almond dragged Reading unconscious from the plane and put his life jacket on. The two men tied their life preservers together and drifted for some time. Then Almond said he felt something hit his foot. Reading told Almond to climb on his back and hold his foot out of the water—and then saw the sharks. Almond showed Reading his legs: his right leg was bitten "all over" and his left had a great deal of the thigh missing. Every time the sharks struck, the two men were badly jerked or were even pulled under the water. Finally Almond was pulled so hard the linking cord snapped and Reading suffered the horror of watching his companion bobbing up and down as he was eaten. Reading was untouched by the sharks, although he could feel them underfoot right up until his rescue a few hours later.

Of the world's 250 species of sharks only about 10 percent are proven man-eaters and, even among these, man-eating can be regarded as exceptional rather than general behavior. In spite of this, sharks pose a problem quite out of proportion to the amount of damage they do. A single shark incident off a popular beach can economically cripple a small holiday resort. In 1958 resort after resort along the normally crowded Natal South Coast in South Africa was quiet and almost deserted. Several people had been attacked by sharks. Hotels were facing ruin and the beaches, warm and inviting, were empty. This short, sharp series of attacks cost the South Coast millions. But economics is not the only problem: the presence of sharks makes several maritime occupations in some parts of the world either hazardous or impossible. In wartime, morale can be seriously affected among men operating in known shark areas. There has always been the problem of ships being wrecked or aircraft having to ditch in shark-infested seas.

There are probably in the region of one hundred shark attacks reported every year from various parts of the world, of which half or more are fatal. But this figure is misleading. What is important in assessing the toll is the number of people who go missing at sea or from coastal areas every year.

In 1966 an experienced skin diver disappeared off the Zululand coast. He went down on the record as "missing." A month or two later another skin diver disappeared a little farther up the coast. He too is still "missing." Knowing this area it seems certain that they were both taken by sharks, but the records can never assume this. Farther north, along the wild and roadless African coast, African and Indian fishermen disappear frequently in their small craft or when wading in the shallows. How many more vanish from among the tens of thousands in the Indo-Pacific area, where, in order to eat or earn a living, the people have to take tremendous risks in the shark-infested sea? Sharks probably kill at least a thousand people a year and perhaps even double that number.

During the last war the losses on both sides through the depredations of sharks must have totaled tens of thousands. Hundreds of ships and smallar craft were sunk and in many cases there were no survivors. But among those who did survive there were some horrifying tales to tell.

The *Nova Scotia* was torpedoed by the Germans at 9:30 A.M. on November 28, 1942, thirty miles from the Zululand coast. There were 900 men aboard, of which 765 were Italian prisoners bound for prison camp. One of the guards, George Kennaugh of Johannesburg, told his story to the late Dr. David

Davies, the Durban shark expert. He said that he was off duty and had changed into a bathing costume for a swim in the ship's pool when there was a tremendous explosion. The ship, almost immediately, began to sink and he fell overboard without his life jacket. "There were hundreds of men around me in the water swimming and clinging to bits of wreckage and rafts. Another South African swam over and clung to the oar. [Kennaugh was using it to keep himself afloat.] He was wearing a life jacket. The two of us drifted on a strong current until next morning. Other survivors were visible around about on rafts, gratings etc. When it became light my companion said it was better to die than to go on holding on like this. He said he was going to let go and refused to listen when I told him not to give up. So I asked him to leave me his life jacket.

"As he was loosening his life jacket he suddenly screamed and the upper part of his body rose out of the water. He fell back and I saw the water had become red with blood and that his feet had been bitten off. At this moment I saw the grey form of a shark swimming excitedly round and I paddled away as fast as I could. Then a number of sharks congregated around me—I estimated their lengths at between six and seven feet. Every now and then one would come straight for me—I splashed hard and this seemed to drive them away."

Kennaugh was dragged aboard a raft which was continually circled by sharks, "which we hit with spars to keep them away." When a Portuguese sloop rescued them the rescuers had to club the sharks away with boat hooks. Out of 900 men, only 192 survived.

Even in peacetime a surprising number of shipwrecked people and downed air crew are killed by sharks. On November 16, 1959, 120 miles southwest of New Orleans, all the crew and forty-two passengers from an airliner which apparently crash landed intact on the sea were drowned or eaten by sharks. Ten half-eaten bodies were recovered by rescuers, who arrived on the scene too late. One hundred and four men took to life belts when the British patrol boat *Valerian* sank in October

1926 off Bermuda, which is not noted for its sharks. But a large number of sharks moved in and began snapping at the survivors. Only twenty were rescued.

Sharks do not always prefer taking live prey. Jean Campbell Butler tells a story about a crew of three Ecuadorian airmen ditched in the Pacific several miles out from the Ecuador coast late one afternoon in 1941. The three men all had life jackets and hung on to each other for company during darkness. During the night one of the men died and the flight officer decided to take at least his body home. He began pushing the man in front of him and then swimming to catch up—all the time traveling toward the coast. Suddenly, in the darkness, the body was snatched away and was never seen again. The horror of the situation told upon the flight officer's second companion and a few hours later he too died. Again the officer pushed the body ahead of him but not too far, for in the moonlight he had caught sight of dark shapes sliding through the sea and making swift passes at him. The skipper then noticed that parts of the corpse's legs were missing but still he stayed with the body. Then he felt the sharks bumping him in their eagerness to get the floating man and in his terror he let go his companion, who was immediately carried off. The flight officer found himself treading on the backs of the sharks at times. Eventually he reached the shore, but the sharks had never left him. Nor had they harmed him.

There are many stories of men who survived by shouting at the attacking sharks or by banging with cupped hands on the surface as they approached. One of the best repellents seems to be a good punch on the nose. Hans Haas, Cousteau, and even the insuppressible Pliny advised that sharks can't take the bold approach.

Another characteristic about shark attacks is that when attacking coastal resorts they invariably go for a solitary bather rather than a group. Once they have attacked a man they will persistently go for him even to the extent of brushing past rescuers and rubbing off great pieces of their flesh with their sandpaper-like skin. Contrary to popular belief, they do not appear to get excited when attacking. They proceed slowly, doggedly almost, and at times even bullets have failed to put a shark off his meal. Few people who go out to rescue shark victims are themselves attacked.

On April 4, 1958, a man named Badenhorst gave me a firsthand account of how he, the day before, had lost his brother, policeman N. F. Badenhorst, aged twenty-nine, who had been attacked by a shark at Port Edward, on the Natal South Coast. "I was swimming near my brother when he shouted 'Shark!' and I saw him kicking at something in the water. The water immediately turned red and I saw my brother's leg was practically off above the knee. I swam toward him. He punched at the shark and it took his arm off. He then rolled over and punched with his other fist and it took that arm off too. Just then a Zulu swam past me and grabbed my brother and carried him to the beach where he died."

It seems natural enough for a shark to assume that anything floating on the surface is edible. Any large pelagic shark (deep-sea sharks which generally feed near or on the surface) will attack a swimming man and various types of sharks have deliberately bumped boats and tried to capsize rubber lifeboats. On one occasion a shark leaped into a rubber dinghy and bit an occupant, who died four hours later.

With coastal sharks there are so many anomalies, so many theories, and so many missing pieces of information that it is difficult to distinguish any orderly pattern. Some species are particularly dangerous in one region and not in another. The Zambesi shark, for instance (also called the bull shark and Lake Nicaragua shark), is the only proven killer in South African coastal waters. It is merely noted as an aggressive shark elsewhere, with few positive kills to its name. On the other hand, the tiger shark is the number one coastal killer in Australia but has no confirmed kills on the South African coast, where it is a frequent visitor.

Patterns of behavior are all important to biologists making a study of a particular animal. But sharks remain inscrutable. When the glimmer of a pattern appears to be emerging it is very often shattered, or at least blurred, by new and puzzling facts. From the facts, it is evident that with very few exceptions sharks do not attack bathers in water that is cooler than 70 degrees. In Australia and South Africa there are no exceptions to this rule. Is this because sharks are not hungry when the water is cooler? Or is it because there is nobody in the surf to eat? Bathers in the sunny southern hemisphere consider the sea too cold for a swim when it drops down into the 60s. Butler records some interesting laboratory tests in which sharks stopped eating for two months when the temperature dropped to 62 degrees. When it had been 71 degrees for ten days they ate avidly. When the temperature came down again they stopped. The world's most southerly recorded shark attack upon a bather was off Tasmania in January 1959 when a naval rating was killed by a shark. This was at latitude 43 degrees south, which has a mean sea temperature in January of 63 degrees—but that particular year there was a heat wave and the sea was at least 8 degrees warmer.

There is the suggestion of a pattern in the fact that out of 433 cases investigated by the Smithsonian Institution, 16.8 percent of the victims were gutting or carrying speared fish. Sharks undoubtedly have a keen sense of smell. Although they appear not to be attracted by the smell of human sweat or urine, the smell of blood, both of fish and of man, affects them. In South Africa there is also evidence that sharks are attracted by dirty water. Ten out of twelve shark attacks in South Africa recently were in dirty water, caused by flooding rivers. This is not borne out in all other "risk regions" around the world, and it might be that just the Zambesi shark is particularly attracted by rivers in flood. In fact these sharks have a definite personal liking for rivers generally and have been found 340 miles inland. Far upstream on the Zambesi they have attacked canoes and turned them over.

Most shark attacks in the world take place between 2 P.M. and 6 P.M., but then that is the time when most people swim. Most shark attacks take place in water two to five feet deep, but few people venture out deeper than that. Nevertheless it serves as a warning to people bathing in shark areas that sharks have attacked in water only two feet deep. There are instances of sharks leaping out of the sea, even onto beaches, to take a bite out of somebody. Nor is the size of the shark any indication that it will or will not attack: a fourteen-year-old schoolgirl was standing thigh-deep in the sea on the Natal South Coast when a four-foot shark pulled her down and snapped an arm off. Even two-foot sharks have inflicted some dreadful wounds, and at least one bather, in the United States, was crippled by one which struck him in knee-deep water.

Butler offers a very acceptable theory that rogue sharks are responsible for many if not most of the worst coastal attacks. In other words, individual coastal sharks become man-eaters just like lions or tigers. The theory is bolstered by the fact that sharks treat fish as their staple diet and most of them are perfectly capable of catching what they need. For them to rush into the shallows and grab a man is exceptional behavior unless they take the sight of a person's leg as a fish (a possibility in murky water) or unless they mistake bathers as creatures in the throes of death; remember that they would only see the lower portion of the body. Cousteau noted that sharks will go for a man on the surface, but when he dives below the surface and swims they usually lose interest.

The rogue theory is bolstered by an impressive amount of evidence. In 1922 off Coogee, near Sydney, two men were taken by sharks within a mile and within a month. Over the next three years two more were taken in the small area—and then no more. Australia's famous handkerchief-size beach, Bondi, in Sydney, has been free of shark attacks in this century except for one year (1928) when four people were attacked. Sometimes a rogue shark appears, in Butler's words, to "carve a path of death." Along a sixty-mile stretch of the New Jersey coast in July 1916, a great white shark attacked five swimmers in ten days, killing four of them. It was later caught and found to be a "mass of human flesh and bone" inside. When it was caught all attacks ceased. The fish was eight and a half feet long and was caught off South Amboy.

On December 18, 1957, sixteen-year-old Robert Wherley went for a late-afternoon swim fifty yards from the beach of Karridene on the Natal South Coast. He suddenly felt a sharp pain in his leg and swam for shore. His leg had been severed at the knee. Two days later fifteen-year-old Allan Green was swimming thirty yards from the Uvongo beach (sixty miles south of Karridene) when he was attacked by a shark. He died immediately from multiple injuries. Three days later and five miles south, at Margate, twenty-three-year-old Vernon Berry was standing in three feet of clear water. It was again late in the afternoon, and there were plenty of swimmers about. He was suddenly seized by a shark, which bit through his left leg. As he fell the shark took off his left hand and then before people could drag him clear, his right arm was terribly mauled. He died on the beach. On December 30, Julia Painting, aged fourteen, was standing in three feet of water a few yards from where Berry had been taken, when she too was dragged down. The shark took off her right arm and mauled her about the body. On the beach the girl begged her helpers to let her die but in fact she survived. A wave of horror swept South Africa and thousands of Christmas holiday makers cut short their holidays and went home. Nine days passed and the tension eased. But on January 9, forty-two-year-old Derryk Prinsloo was attacked fifty miles north of Margate, at Scottburgh, as he floated in shallow water ten yards from shore. The water was murky and rough when the shark took out great pieces from his buttocks and legs. He died in the water and was brought ashore with his left arm almost severed. Almost three months went by before N. F. Badenhorst was taken seventy miles south at Port Edward, where he had been swimming in shallow water in the early afternoon. He lost three limbs and was brought ashore dead. Two days later and twenty miles north—in fact back to the scene of the second attack—Mrs. Fay Bester, aged twenty-eight, was in four feet of murky water twenty yards from the shore of Uvongo beach when a shark grabbed her around the middle and shook her like a dog shaking a slipper. Mortally injured, she struggled briefly before the shark bit almost clean through her left leg. Thus inside four months and within a hundred miles of coast seven people had been attacked, of whom five had died. Was it just one shark? It would be incredible if it were not. It was almost a year before there was another shark attack along that coast. A fifteen-year-old had his left leg taken off by a shark at Port Shepstone (he survived) and then almost two years later a sixteen-year-old youth was badly lacerated fifty miles north; but he too survived. That same year (1960) Petrus Sithole, aged twenty-five, was swimming off Margate's beach when he lost both legs and died. During the years that followed there were several more attacks including three in the space of two months which were fatal. In one of these a middle-aged man was brought ashore with only his head, forearms and lower legs unstripped of flesh. It is possible that he had already drowned when the shark or sharks attacked him.

A lot of people who might otherwise have died are now surviving even the most dreadful injuries because of new techniques in the treatment of shark-attack victims. No longer is it deemed advisable to rush them to a hospital: victims are now (in most countries) left on the beach for half an hour with their head pointing down the slope and tourniquets applied. They are no longer covered with heavy blankets and where possible (in South Africa at any rate) a doctor is there before the patient is moved to a hospital. A quarter of a grain of morphine and intravenous shock therapy is automatically given. In the past, most people taken out of the sea alive have died from either shock or loss of blood and only rarely because a vital organ had been injured. Eighty-seven percent of people attacked are injured below the waist. Most victims are bitten at least two or three times—only 10 percent are bitten only once. There is evidence too that most are "mouthed" or scraped by the shark's extremely coarse skin before being bitten.

There is hardly an area in the world which can be said to be free of the danger of shark attack, although naturally in some areas the risk is so slight as to be negligible. The high-risk zone can be taken as a belt between 23½ degrees north and 23½ degrees south. In this tropical belt there is always the risk of being attacked by sharks. The risk zone expands north to 40 degrees in June, July, and August and then south in December, January, and February. Although pelagic sharks might attack anywhere, the most northerly coastal shark attack was probably one at Wick on June 27, 1960, when Hans Yoachim Schepper was mauled. Sharks—possibly tiger sharks, which are found right up to Iceland—have frequently attacked small craft in the English Channel, the Irish Sea, and the North Sea. The most northerly seaside attack in the United States was in July 1936, when sixteen-year-old Joseph Troy died after his leg was almost severed in Buzzards Bay, Massachusetts (41 degrees, 38 minutes north). Next to Australia, the United States appears to have the second worst record for shark attacks. South America has a few and would possibly have more if it had more resorts.

Parts of the Pacific are notorious for shark attacks, but Hawaii's coral-protected beaches are fairly safe—although when an aircraft crashed off Maui on February 3, 1961, three of four passengers were eaten.

The record in Australia is fairly bad, especially on its east coast (probably because of a warm current), and New Zealand, although only half in the risk zone, has suffered three fatal attacks in this century, one of them in January 1966 off New Plymouth.

The Far East records several attacks each year, although hundreds must go unrecorded—which is also the case in India, where the mouth of the Ganges is apparently a favorite spot for man-eaters. There was a time in India when pilgrims, overcome with a fervor to sacrifice themselves, waded into the sea so that the sharks could take them. According to David Kenyon Webster the sharks did just this until the sea was bright red and finally the sharks were so gorged they could eat no more and many were the disappointed uneaten pilgrims.

In the Middle East, especially in the Red Sea and the

Persian Gulf, there are attacks each year, and over twenty-four years there have been twelve attacks one hundred miles up the Karun River near Ahwaz in Iran. The Mediterranean Sea was said to have been free of man-eating sharks until the Suez Canal let them in. Today, states Butler, it is one of the worst areas in the northern hemisphere. Attacks have been recorded with fair frequency off Israel, Egypt, Yugoslavia, Greece, the Bay of Monaco, Genoa, and Venice.

The shark as a man-eater has the distinction of being one of the very few on which man gets his own back, for man is a great shark-eater. There is a tremendous amount of food value in sharks (although some of their livers are toxic). One could fill a cookbook with shark dishes, some of which are highly acceptable even to a conservative palate. The time could come when sharks will be slaughtered on a wider scale for human consumption.

The two largest sharks, the whale shark, which grows to sixty feet, and the basking shark, are perfectly harmless and feed only on plankton. A great many of the other two hundred or so species of shark are dogfish. Their ancestors emerged at the start of the Carboniferous Period 275 million years ago and they have developed into perfectly streamlined cartilaginous fish (no bone skeleton) which are magnificently equipped for their environment. Once out of the water, sharks fall into a shapeless mass of flesh.

There is a great deal of controversy over which of the score or so of man-eating sharks are the most dangerous. There are three that appear to cause a great deal more concern than other dangerous sharks:

The blue pointer (*Carcharodon carcharias*)—also called the man-eater, great white shark, white shark, death shark, and white death shark—this notorious animal is a descendant of the extinct ninety-foot *Carcharodon megalodon,* which had teeth four inches long. The blue pointer has large teeth with serrated edges like steak knives. Since these sharks grow to thirty feet and more, they are easily capable of snapping a man cleanly in half. J. L. B. Smith claims they can swallow a man whole and describes them as "the most feared and ferocious of all sharks." Dr. David Davies called them "perhaps the most dangerous of all sharks." This shark is not common, but it still chalks up an enormous toll of life. It usually hunts alone and is responsible for several deep-sea and coastal attacks, including a number in which boats have been attacked. Most of the attacks outside the high-risk zone are the work of this shark and so are most of the fatal attacks at the northern extremities of the West Coast and the East Coast of the United States.

Tiger shark (*Galeocerdo cuvieri*): this is a widely distributed shark which again is large enough to bite a man in half; in fact, J. L. B. Smith claims it has done just that. Coppleston and Whitley say it is probably the worst killer along the Australian coast and there are reliable accounts of its killing men in the West Indies and off South America. It has frequently attacked boats. When a tiger shark becomes big and old it may become sluggish and prefer to look for carrion, but young ones are extremely lively. J. L. B. Smith describes it as "one of the most fierce and dangerous of all creatures." A fourteen-footer taken at Durban had in its stomach the head and forequarters of a crocodile, the hind leg of a sheep, three seagulls, two two-pound tins of peas (unopened) and a cigarette tin. One caught off Florida had the remains of two men but it was never established whether the men were dead or alive when eaten.

The Zambesi shark (*Carcharhinus leucas* spp.), also known as the bull shark, has near relatives such as the fresh-water Lake Nicaragua shark and the Ganges shark, which have similar habits. The Zambesi shark has killed men well inland on the Zambesi River and is responsible for most deaths off South Africa, just as the Ganges shark has accounted for many off India. Under the name of bull shark it is one of the more aggressive sharks in American waters. Although it has serrated teeth like the tiger shark, it does not amputate as cleanly; it usually gives a vigorous shake first.

The whaler shark (*Carcharhinus macrurus*), which might

be identical to the Zambesi shark, has been blamed for most harbor attacks in Australia. Its dentition is only slightly different from the Zambesi shark, and it has the same habit of shaking its victims.

Other proven man-eaters are the mako (*Isurus sp.*) and the hammerhead (*Sphyrna sp.*), while the following are considered probable man-eaters: the blue shark (*Prionace glauca*), the lemon shark (*Negaprion brevirostris*), night shark (*Hypoprion signatus*), ragged-tooth shark, Atlantic sand shark (*Carcharias taurus*), porbeagle (*Lamna nasus*), thresher shark (*Alopias sp.*) silky shark (*Carcharhinus flaciformis*), blackfin shark (*Carcharhinus limbatus*), white-tipped shark (*Carcharhinus longimanus*), black-tipped shark (*Carcharhinus maculipinnis*), dusky shark (*Carcharhinus obscurus*), sandbar shark (*Carcharhinus milberti*), springer's shark (*Carcharhinus springeri*), nurse shark (*Carcharhinus arenarius*), and the reef shark (*Carcharhinus menisorrah*). There are one or two smaller species of shark which are known to have taken a bite of a man on rare occasions. Since they are not big enough to kill a man, they need not be listed here.

The thought of men being attacked by sharks is horrific enough for millions of dollars to be spent on research into repellents. It might be unduly pessimistic to say all shark repellents—except shark nets—have failed miserably but it would not be far from the truth. Electrical gadgets, chemicals, sonar equipment, compressed air, sharpshooters—one after another they have proved unsuccessful. A Hollywood film producer suggested at a press conference that dolphins could be used, for they will drive away sharks. He showed a film of dolphins attacking sharks to save the life of a small boy to prove his point. The idea received widespread publicity, but a few weeks later, talking to his main cameraman, I was told: "Hell, those weren't real sharks, we had to use rubber ones. Sharks would have cleaned up that dolphin in no time!"

In Paris in 1965 a life belt underhung with a stout canvas bag was exhibited as a possible antishark device for shipwrecked people. Since sharks have bitten through fiber glass boats as well as rubber rafts, the canvas life belt does not fill one with confidence.

The most popular shark repellent—in vogue in many parts of the world for twenty years—is the "Shark Chaser," which was developed for wartime personnel operating in shark areas. It is a six-and-a-half-ounce tablet containing 20 percent copper acetate and 80 percent nigrosine-type dye, which stains the water a bright orange and can be seen from spotter aircraft quite easily. In spite of all claims for it, however, sharks have sometimes taken no notice of it. In one incident recorded by J. L. B. Smith a shark rushed up to a fish which was nibbling at a floating repellent tablet. The shark ate the fish. Then it headed for the tablet, which it swallowed. A few moments later the shark rejected it.

Copper acetate seems to be of no use against hungry sharks. During the war, when morale was undermined by the shark threat, the United States Army authorities used the only resource they could: they made fun of the shark and tried hard to debunk the shark threat by issuing a little survival book filled with funny pictures.

If there is one thing that almost always deters a shark it is men together in a group. During the war when rescuers came upon groups of men hanging on to crowded rafts they noticed that the sharks would make only furtive passes at the men in the water. Once a man was bitten they would keep snapping at him until he drifted apart from the rest; then they would attack him furiously. But when the group broke up to swim to the rescue boats the sharks became bolder and took bites where they could. Sharks have often been described as cowardly, but their reluctance to attack a man who offers effective resistance is easy to understand, for a shark can pass its entire life and not meet a sea creature that will fight for its life quite as desperately as a man.

In June 1958 the American Institute of Biological Sciences established its Shark Research Panel under the chairmanship of the zoologist Dr. Perry W. Gilbert. The panel was charged with collating, on a world basis, every atom of fact on sharks and shark attacks and to find out any patterns that exist. If Gilbert had any misgivings about his task, they were justifiable. The panel found so many apparently unrelated and contradictory facts and so many imponderables about sharks and shark attacks that they were in the same position as the Oriental potentate who was pushed into a strange and overcrowded harem. Although he knew precisely what was required of him he hardly knew where to begin.

Captain Jacques-Yves Cousteau, the pioneer underwater explorer, confessed once: "We never did get to understand sharks." These words might well have been repeated by the thirty-four delegates after the 1958 New Orleans Shark Conference. The pick of the world's shark experts had been there to discuss the possibilities of developing a really effective shark repellent and had delivered a wondrous selection of papers on sharks. The only significant conclusion that could be drawn from the conference was that there were a prodigious number of gaps in man's knowledge of sharks—the last undefeated man-eaters left in the world. In 1958 sharks were indisputably on top. Today they still are.

VIII
FLYING KILLERS

The Vulture

Birds

"What does the little birdie say
In her nest at break of day?"
Alfred Tennyson

Early in 1967 the following story was put out by an international press agency: "Teheran.—The newspaper 'Ettelaat' said yesterday two children, aged five and three, were killed when they were dropped from the sky by two giant eagles which swooped on them in the village of Jegelan in north-west Iran. The paper's correspondent said the children's father had earlier removed eaglets from the birds' eyrie." This story, almost certainly untrue, was probably read in most countries in the world and confirmed what most people have always believed: that in certain areas giant eagles snatch up children and carry them off. I am skeptical because there is not a bird with both the wing strength and the talon strength to lift a human off the ground. The two species of birds which are usually blamed for child-snatching are the lammergeier, *Gypaëtus barbatus,* and the condor, *Vultur gryphus,* both of which are vultures which prefer carrion or small live prey.

But even smaller birds of prey have been blamed. Norwegians will swear that golden eagles in Scandinavia carry off children, and I have heard Yugoslavs assuring each other of similar supposedly authentic cases. The stories were born of ancient folklore and nurtured by rumors. Osmond P. Breland states categorically: "I do not believe there is a single verified case." Neither do I. Hypothetically a large bird of prey such as the lammergeier can carry off a human infant, for there are instances of them lifting animals weighing up to ten and fifteen pounds.

There is but one fairly reliable case of a bird of prey killing a man. The victim was the Greek tragic poet Aeschylus, who, at the age of sixty-nine (in 456 B.C.) was killed instantly when a lammergeier dropped a tortoise on his head. The bird was obviously trying to break the tortoise's shell so that it could eat the contents. For a man who had survived the battle of Marathon and had written eighty plays it was an ignominious end. At least, it was probably unique.

In October 1966, a report from Trento in Italy said a housewife "killed a huge eagle with a rake after it tried to snatch her three-year-old daughter." On October 20, 1966, the Sketch News Service issued the following report of the same incident:

"Rome, Thursday--a housewife has told of a battle in which she killed a royal eagle. The bird tried to fly off with her three-year-old daughter.

"For three minutes, Mrs. Angela Vidola (37), fought the enormous bird with a kitchen mop in the back garden after it grabbed her child by the pullover.

"Mrs. Vidola, who was preparing supper when the eagle swooped, said: 'Suddenly I heard a furious flapping of wings and the squealing of three terrified dogs. I ran into the garden and saw an enormous eagle. It had already killed three chickens and was trying to take off with a turkey.

" 'It suddenly dropped the turkey and made for my daughter Flavia. It hovered over her, flapping its enormous wings, then caught her blue pullover.

" 'I shouted to Flavia to lie on the ground. I swung the mop again and again and managed to hit its head many times: it fell dazed and I kept hitting and hitting it. . . .'

"Mrs. Vidola, who was badly scratched, then collapsed."

I think the assumption that the eagle was attempting to "fly off with her three-year-old daughter" is unjustified. From subsequent investigations it seems that the eagle did attack the child—probably in panic—and its talons became ensnared in the pullover. It would have been totally unable to lift the child. From such alarming incidents, obviously, the more dramatic ones are born. Reynard Oberholster, an authority on the wildlife of the early days of the Transvaal, told me a similar story involving Oberholzer of Oberholzerskloof in the Marico district. Oberholzer was cycling home one day in 1923 when he heard a strange rushing noise behind him. Then he felt an eagle grab him behind the neck. He fell off his bike and fought and killed the eagle. Why did it attack him? Did it perhaps misjudge his size and take him for a fleeing animal?

There are several more stories about eagles and owls of various species flying at people, but apart from the odd defensive gestures when the bird's nest is approached, the incidents are difficult to explain. Even birds the size of sparrows will defend their nests. There is a slight possibility that a determined bird of prey could kill a man; it would be foolish to say it could never happen. Some eagles for instance have been seen to fly into the faces of comparatively large animals such as antelope and drive them backward over cliffs. Then they eat them. It is conceivable that one has in the past killed a man in this way.

The roc or rukh which carried off Sinbad, and to which Marco Polo referred, never existed. The bird was supposed to have been powerful enough to lift an elephant and many gullible potentates accepted specimens of its enormous feathers from travelers. The "feathers" were raffia palm fronds, which resemble enormous quills. So far as is known, there has never been a flying animal big enough to carry an adult man. Even the enormous *Pteranodon ingens,* a flying reptile with a wingspan of twenty-seven feet which existed in Kansas ninety million years ago, could have possessed nowhere near the strength to carry a human adult. There were, of course, no humans on earth at that time anyway.

Several flightless birds have been capable of man-killing. The biggest of these so-called ratites was the now extinct elephant bird (*Aepyornis*) of Malagasy, which weighed just short of half a ton (965 pounds). It stood nine to ten feet high and laid eggs the size of a two-gallon bucket. The bird became extinct within historical times and preserved specimens of its eggs probably gave rise to the legend of the roc.

Within the last two hundred years lived *Dinornis maximus,* the New Zealand moa, which undoubtedly was frequently forced to kill Maoris who hunted it to extinction. The New Zealanders fought and killed the massive birds, using the most crude wooden weapons, and a thirteen-foot-high bird (they stood almost as high as a double-decker bus) weighing up to 520 pounds would certainly be capable of killing a man with a single kick or even a jab with its heavy bill. W. R. B. Oliver states: "The moa was said to be able to run with great speed and to be able to defend itself by kicking. Both dogs and men were reported to have been killed by a kick from a moa."

Of the extant ratites two are said to be capable of killing men. One of them, the ostrich of Africa, is bred extensively in parts of the Cape Province, where from time to time it kills farmworkers and farmers. Its forward kick—it is incapable of kicking backward—has been compared with the kick of a mustang. When the birds kick, a sharp, clawlike toenail flicks forward like a switchblade and with this the ostrich has disemboweled men. However, most fatal injuries inflicted on ostrich-farm workers are to the head and I can only imagine this is because the bird leaps into the air as it kicks forward. A really big bird can stand eight feet high and weigh 250 pounds and one monstrous bird tipped 345 pounds and stood nine feet high.

One can describe the pugnacious ostrich as the world's most dangerously aggressive bird, and down among the ostrich farms in the Oudtshoorn district I found that the longer men knew ostriches the more respect they had for them. Few moved without a branch of *Acacia karoo,* which has vicious, stiletto-like thorns, with which to jab at the ostrich should it attack. There seem to be two or three serious incidents a year down there; the latest fatality, in 1965, involved Dirk Avontuur, a twenty-year-old of Le Roux, who died several months after an ostrich cracked his skull.

In the wilds the ostrich can be just as pugnacious should one approach its nest and such vigorous defensive behavior belies the Bible's observation: "She is hardened against her young ones, as though they were not hers" (Job, 39:16). Men who have survived an ostrich attack advise one to lie face down on the ground, using the hands to protect the back of the neck against the ostrich's pecking. In this position one is relatively safe from being disemboweled by a kick. It is no use getting up and running, for ostriches can run at 35 m.p.h.

The ostrich's near relative, the Australian cassowary, is also said to be able to kick a man by jumping in the air and lashing forward with its heavy, stiletto-armed feet. It is said to have killed one or two people.

Dramatic as these large birds might be, it is the little birds, such innocents as budgerigars, and even hens, which are the greatest man-killers.

The medical world never really appreciated the danger until in 1929 the annual caged-bird show ended in Córdoba in Argentina. The show had been notable for the number of parrots that died. It was also marred by one or two human deaths from pneumonia. No connection was seen until the following year, when many of the birds were exhibited at Tucumán in northern Argentina. Again a number of parrots died and the organizing committee held an inquiry. During the inquiry they became aware of the number of exhibitors' deaths—again from a form of pneumonia. There was obviously a link, but before anything could be done to check its spread birds from the exhibition had already been exported to the United States, Europe, and Africa. It was not long before the disease flared up in those parts. In all, four hundred cases were reported, of which more than a third of the victims died.

The disease is now known as psittacosis and is caused by the virus *Miyagawanella psittaci*. It is carried mainly by the parrot family, Psittacidae. A parrot can transmit the disease to man by pecking him, or man can pick it up from the dust while cleaning the bird's quarters. There are a few hundred cases reported every year and quite a number are fatal.

Parrots and budgerigars are no longer considered the only hosts of the virus. Some seventy species of birds are thought to be carriers and nearly half the cases reported in the late 1950s were caused by turkeys. Fifty-five percent were due to Psittacidae.

Because of psittacosis the importation of budgerigars was stopped in Britain in 1930. In 1952 the banning was temporarily lifted and in that same year, in a Birmingham store, a newly imported parrot pecked a woman lunch-time shopper. She died shortly thereafter. The ban was quickly reimposed.

Apparently the disease's symptoms are initially similar to those of typhoid and the younger a person is the more chance he or she has of recovering.

Another serious problem presented by birds—albeit indirectly—is the problem of migrants transmitting disease-carrying ticks. Thus African tick-borne diseases have broken out in Russia and India, and ground-feeding birds appear to be the culprits. An excellent example is the Kyasanur Forest disease, which is fatal in one out of ten cases. The disease was first noticed in India in the northern spring of 1957 in the Shimoga district of Mysore State, where large numbers of monkeys were dying. There had been a number of human deaths in the area resulting from a mysterious fever. The Virus Research Institute in Poona, India, investigated, at first believing it to be an outbreak of jungle yellow fever. Whatever the Kyasanur Forest disease is, it is closely allied to this disease. According to a WHO report the Poona Institute quickly established the virus carrier as the tick *Haemaphysilis spinigera* and found plenty of the ticks on the Kyasanur Forest floor (hence the name). The problem was: how did these hitherto unknown ticks get there? The parasitologists collected ticks from monkeys, small mammals, and birds and as the disease was exotic the only possible culprits were birds (ticks cannot fly). WHO supported a scheme carried out by the Bombay Natural History Society to investigate thoroughly all possible bird hosts and their flyways. To date 10 percent of all birds banded have been tick-infested but it will be some time before their role in disseminating the disease becomes clear.

Bird flocks of unbelievable numbers are creating another kind of problem in Africa. Swarms of seed-eating quelea finches (*Quelea quelea*), a tiny, handsome bird, have destroyed hundreds of square miles of grain crops in central and southern Africa in recent years. In some areas they have created a high mortality rate resulting from malnutrition. During the 1960s southern Africa suffered a devastating drought which put whole tribes on the brink of starvation—then came the queleas to aggravate the situation. In one raid on the Springbok Flats by the South African government, using a contact poison, twenty-five million birds died.

Today in this part of Africa the quelea finch has eclipsed the locust as an agricultural problem. In fact, it is called the locust bird. Every year tens of millions are killed.

These birds have become a serious nuisance only in recent years for three reasons: (1) a prolonged drought destroyed much of the veld, forcing the birds to seek food on developed land; (2) less and less undeveloped land is available for them; (3) and most important, new irrigation schemes and dams have allowed the birds to travel around in search of food. The quelea finch is a thirsty bird and must have a good supply of water as well as grain to survive. So vast are the breeding swarms that when they swoop down at dusk to skim the water surface to drink, thousands are drowned by the pressure from the flock above. Boughs of trees snap off under the weight of roosting birds and as the alarmed queleas rise momentarily into the air they resemble, from a distance, a wisp of rising smoke.

With the world's human population at more than three billion and the world's bird population, estimated by James Fisher, at about a hundred billion, it is natural enough that there should be growing conflict between the two. Now there is a new and growing menace presented by the birds: midair collisions between them and supersonic aircraft. A tiny swallow, being struck by a supersonic jet, could tear through it like a piece of shrapnel.

Midair bird-plane collisions cost the Royal Air Force a million pounds a year (nearly two and a half million dollars). Recently in the United States 300 airliners and 530 military and private aircraft were damaged by birds in a single year. There were three deaths.

One of the worst disasters caused by birds occurred in Boston in 1960 when an Electra turbo-prop struck a cloud of starlings. The collision made both engines cut out. The airliner nose-dived into the water, killing all sixty-one passengers and crew.

It is logical to suppose that as airports become busier the incidence will increase. With the advent of ultrafast and massive jetliners, the problem is likely to be given a great deal more attention.

Bats

"Twinkle, twinkle, little bat!
How I wonder what you're at!"
Lewis Carroll

We were sitting around the card table playing a quiet game of canasta in the Swaziland highlands. Outside, in the forest clearing, the shrill-voiced frogs and swizzling crickets had begun their night-long chorus. It was the sort of monotonous, unbroken noise that one's ears cease to register after a few minutes and only when they all inexplicably stop and the silence comes in like a percussion wave is one aware of the din they were causing. Through the window I could see the white disk of the moon riding above the trees. It was a pleasant, peaceful night. Suddenly, from the kitchen, came the sound of a cup shattering on the tiled floor. Then came a scream—a frenzied, desperate scream. Our chairs crashed back and we rushed into the kitchen to find the servant—a young Swazi woman—sagging gradually to the floor, her face ashen with terror. Around the kitchen flew the culprit: a bat. We grabbed towels and brooms and began to flail the air until the creature flew into the passage and then into the bathroom, where we finally managed to wing it. It fell in a crumpled, ugly little heap in the bath. Picking it up in a towel, for nobody wanted to actually touch it with his fingers, we tossed the concussed animal into the night.

In retrospect I felt quite embarrassed about the affair and was glad nobody pursued the subject. Even the servant preferred not to talk about it. What is it, I wondered, that makes us so squeamish about bats?

Part of the answer must lie in the appearance of the animal.

The world's enormous variety of bats have heads that are grotesque caricatures of such widely diverse animals as horses, foxes, cats, and rats; some of them look like tiny, hideous, winged humans. The females even have twin breasts in exactly the same position as human breasts. It was this peculiarity that made some zoologists suspect bats were more closely related to monkeys and men than to shrews and rodents. Altogether the bat is looked upon as a sinister and nightmarish animal. Most people think of it as a consort for witches and ghosts rather than as a creature which usefully chases moths or eats fruit.

There is one bat above all others that has helped indelibly to blacken the image of Chiroptera (the order of bats)—the vampire. It is a revolting enough creature in expression as well as in habits and on its head rests the blame for a number of human deaths.

But the blood-drinking vampire is not quite what it is made out to be in popular reading matter: it is not a large, green-eyed, long-fanged animal which taps on windows at night and buries its teeth into people's jugular veins in order to suck out their life's blood.

There are three species of vampires; all belong to the family Desomondontidae, and all are confined to the Americas. They are small, long-eared, rat-headed creatures with short but incredibly sharp incisors and clefted lower lips. They use their incisors to bite a small wound. The blood wells up into the wound, and this they lap up; they do not suck. The small, tube-like stomach, especially adapted for a blood-only diet, cannot take much more than an ounce of blood at a time, which is more or less the vampire's nightly consumption. Of course, if one says an ounce of blood quickly it does not sound much but then imagine a cave full of vampires—say a thousand of them —each needing two large buckets of blood a year. Collectively, says Peterson, a thousand would need 5750 gallons of blood annually. Put another way, the colony would need fifteen gallons a night from the surrounding neighborhood.

Although the vampire has been known to drain a bird of its blood, the amount taken from a human is not harmful.

There is, however, one case of a boy in Trinidad who was host to fourteen bats in a single night. The vampire usually makes its incision gently and undetected in between a sleeping person's toes but will occasionally attack the lips, earlobes, forehead, and fingers. Sometimes a person will wake up to find feet or head soaked in blood.

The danger from vampires in the American Neotropics is not so much in their blood-drinking but in their acting as vectors for rabies. They spread costly rabies epidemics among stock in South and Central America, and on occasion rabid vampires will venture out in daylight and bite animals and sometimes men. Under these circumstances they might even bite other bats, infecting them with rabies, and these bats might then attempt to bite men. I used the word "attempt" because most fruit- and insect-eating bats would find it difficult to bite into a man's skin, and the danger from these bats is not a serious problem. Bat rabies, a variety of the more common form of rabies, is about 9 percent fatal if untreated.

It was not until 1932 that bats were definitely known to be carriers of rabies to humans. A team of experts was called to Trinidad, where a dozen people had died there of a peculiar form of rabies—peculiar because the classic symptom of hydrophobia was lacking. This rabies was characterized by a paralysis which affected the patient just before death and the medical team immediately recognized it as similar to the paralyzing rabies that affects cattle. Each of the victims had been bitten once or even several times by vampires. Vampire-carried rabies still causes a few deaths—very few indeed—in the Neotropics of America and the peasants are not convinced of the danger of these bats. Nor are they particularly repulsed by the idea of having them feed off their blood while they sleep.

The incidence of bat rabies in the Americas is small compared with the incidence of the more common rabies, known sometimes as hydrophobia.

Through recorded history the incidence of hydrophobia has risen and fallen throughout the world with no discernible pattern. In the Middle Ages it was almost epidemic at times and it was fairly commonplace for people to be bitten by rabid wolves. There was a brief decline for two centuries which ended in the eighteenth century, when the incidence once more shot up in Europe. Some authorities say that it was in the mid-nineteenth century that hydrophobia crossed the Atlantic to the New World and it appears to be still spreading in that region today. In 1966 there was some concern in the United States when large wild-animal populations were found to be affected in forty-four states. In Europe, Asia, and Africa there were clear signs in the 1960s that rabies was once again spreading, and it seemed that there is a constant and possibly ineradicable reservoir for the disease among wild animals. For instance, in 1962 in Germany, 2000 foxes and some 2600 other animals, most of which were wild, were found to be rabid. The World Health Organization found that in 1960 there were 611 cases of human rabies reported in the regions under its control. In 1961 there was a dramatic drop to 300 or so. But in 1962 the incidence rocketed to 1453 and nearly half a million people had to be treated for bites by animals that were rabid or suspected to be rabid.

Rabies is a disease of the central nervous system and it can be carried by a tremendous variety of animals, both domestic and wild. Wolves, foxes, coyotes, jackals, squirrels, badgers, certain birds, dogs, horses, goats, cows, and sheep can carry the disease and infect man through their saliva by biting him. Eighty percent of bites are from dogs. The toxicity often depends upon the animal which did the biting: for instance, the rabid bite of a wolf is 80 to 90 percent fatal if untreated, whereas the rabid bite of most dogs is only 16 to 20 percent fatal. The mortality rate is greatly reduced by the use of vaccine—developed by Louis Pasteur in the 1880s—and drops to something less than .5 percent in all types of bites.

The victim of a rabid bite may not show signs of rabies for up to eight months, but once the classic symptom of hydrophobia is manifest the patient usually has less than five days to live. During this period he experiences bouts of violent insanity and may bite another person, thus spreading the infection.

The disease has been eradicated in Great Britain, Denmark, Norway, Sweden, Australia, New Zealand, and Hawaii, where strict quarantine laws guard against its reintroduction.

IX
SMALL BUT DEADLY

P. D. Hugo

The Black Widow

Arachnids

"Along came a spider
And sat down beside her
And frightened Miss Muffet away."
Nursery Rhyme

I can see the scene now in vivid Technicolor. The hero is about to pull on his riding boot when his companion, who has been picking his teeth with a silver toothpick, suddenly leaps forward and cries, "Stop!" He snatches the boot from the hero's hand, turns it upside down, and shakes it. Something falls from it on to the floor. Close-up: a shiny, black widow spider tries to escape. Splat. The companion crushes it with his foot while the audience shudders. The companion, looking down at the nasty little spot on the floor, says grimly, "Somebody must have deliberately put it in your boot . . . somebody tried to murder you!" The hero arches an eyebrow—and well he might. Putting a black widow spider in a man's boot in the hope of killing him is about as efficient a way of murdering a fellow as putting a roller skate on his stairway. The chances are that nothing will happen.

Latrodectus mactans is a shy, secretive, nonaggressive creature no bigger than a little pea, but through the medium of popular fiction it has been libeled as have few other creatures. Looking at what there is of international statistics on deaths through black widow spiders I doubt whether there are more than one or two a year. In Australia *Latrodectus mactans* causes about one death every ten years. In New Zealand *Latrodectus basselti*, known locally as the kapito spider, has killed only three people in this century. In South Africa the two local species *Latrodectus indistinctus and Latrodectus geometricus* have killed only one person—an eight-year-old girl—in postwar years. In the United

States, where the black widow is really notorious, it has killed only fifty-five people since the year 1726. This figure, arrived at by R. W. Thorp and W. D. Woodson, works out at only one death every four years. Altogether Woodson and Thorp traced 1291 attacks during the past two centuries.

B. J. Keyter puts the mortality rate at less than 2 percent for the South African species *Latrodectus indistinctus* (which is about as toxic as the American black widow) and says that 50 percent of people bitten feel little more than a severe local pain with localized swelling. The other 50 percent really suffer, but only 2 or 3 percent have to fight for their lives. A friend who was bitten on the elbow said that within a few hours his arm swelled so suddenly that it split his shirtsleeve. He suffered from nausea and profuse sweating for some hours after.

The spider's venom is, like so many venoms, something of a mystery but in character is similar to a cobra's venom. Weight for weight, it is far more potent. Naturally one receives only a minute dose, but this can be enough to cause quite painful neurotoxic symptoms. The symptoms can last forty-eight hours and in severe cases can be accompanied by acute muscle spasm.

An antiserum is available for the spider's bite and if administered within a few hours is rapidly effective.

Here is yet another case of the female of the species being more deadly than the male. The male black widow spider is a tiny, completely harmless, skeletal creature which, once he has mated with his relatively enormous wife, must run for his life. If she catches him she will eat him.

The tarantula spider or wolf spider, *Lycosa tarantula,* is often accused of being a man-killer and, until the nineteenth century, was blamed for a disease known in Europe as tarantism, which was supposed to be fatal. The only cure for the disease, according to contemporary physicians, was music and wild dancing. The more people who joined in, the better the cure would be. It might well explain why the disease was so popular, for it appears that tarantism was no more than a form of hysteria. The tarantula spider is absolutely harmless to man.

The Australian funnel spider (family *Mygalomorph*), which is reminiscent of the tarantula, has caused the deaths of a two-year-old boy (who died within one and a half hours of being bitten), a five-year-old girl (she died in one and a quarter hours), and two middle-aged women who died several hours after a bite. The funnel spider is normally placid and its bite is only very rarely fatal.

Pliny the Elder, who lived from A.D. 23 to 79, was pretty emphatic about scorpions, which he called *pestis inopportuna* or "a horrible plague." He stated that the "bite" is always fatal to girls and usually to women but the only time it is dangerous to man is in the morning. He warned that they could "fly" on a south wind. The only remedy for somebody who had the bad luck to get in the way of one was a cock's brain in vinegar, five ants in a drink of some sort, or—if all else failed—a little sheep's dung in vinegar applied as an ointment. The Egyptians would have laughed had they heard this advice, for they believed firmly that women were immune to scorpion venom and that scorpions came from the decomposing corpses of crocodiles.

Even now there are some delightfully naïve beliefs concerning scorpions and not a great deal of scientific knowledge.

Depending upon which reference book one believes in, there might be three hundred or five hundred species of scorpions in the world. For our purposes, they can be divided into two types: those whose venom causes only local effects, and those whose venom causes more general and occasionally fatal effects. The former are completely harmless and their sting is about as painful as a wasp's. The common *Euscorpius italicus* of Europe, *Centruroides vittatus* of the United States, and *Centruroides margaritatus* of Central America would number among these. The more dangerous scorpions include two fairly common species in the United States: *Centruroides sculpturatus* and *Centruroides gertachi*.

Scorpions are able to live only in the warmer regions of the world and are found in most of the southern states of

America as well as Central and South America right down to Patagonia. They are common in Africa and fairly common in southern Europe, being found in Greece, Yugoslavia, Italy, Spain, and occasionally as far north as southern Germany. They have never existed in Britain. While it is impossible to gauge how many people are killed annually by scorpions, the number is certainly a few dozen, most of which are in North Africa and the Middle East. During the Second World War, army personnel on both sides during the Western Desert campaign lost a few men who neglected to shake out their boots or bedding—favorite sleeping places for the *buthus,* which have an extremely painful sting.

Like most creatures that have the power to kill a man they have little desire to attack him and have to be provoked into stinging. They are very easy to catch if one catches them from behind the sting using thumb and forefinger. Baboons do this with impunity before ripping off the sting and eating the rest.

The exact nature of the scorpion's venom is not known, but in the case of *Buthus australis* the fluid dries, leaving 25 percent of its volume in crystal form; this can be used for the manufacture of antiserum which is both reliable and effective. The venom, like the black widow spider's, is also similar to that of some snakes. Because of the minute quantity involved it is nowhere near as dangerous and the prognosis is generally excellent. If one is stung by a really large scorpion—they have been found up to seven inches—the results can be fatal in a fairly short time if untreated. The effects begin with a sharp pain in the region of the sting followed by numbness in the limb. Speech then becomes slurred, and one loses control of the salivary glands and often of the bowels and bladder. Breathing becomes labored, and at this stage it is very difficult to reverse the effects. If a person has the ill fortune to be stung directly into a vein, death can come in moments. In December 1966 a twelve-year-old girl in Palapye, Botswana, was stung in a vein and died within five minutes.

If one were to write a poem on the death of an eighteenth-century mountain man or fur trapper in the Rockies one would have him go down with an arrow in his back or at least underneath the weight of a decent-size bear. But in fact few of them died such dramatic deaths and more were killed by the little gray bladderlike tick which sucked the blood of sheep and men with fine impartiality. The tick is the vector of a disease known as Rocky Mountain spotted fever which, even today is 18 percent fatal in untreated cases throughout its North American range. The disease, recognized since it was diagnosed on the Snake River in Idaho late in the last century, is caused by a micro-organism passed into the blood by the tick which is its principal host. The micro-organism, which is larger than a virus but smaller than a bacterium, is known as *Rickettsia rickettsia* and is carried in several species of ticks.

The disease manifests itself in red spots, fever, and pain in the joints and can end in delirium and death. Although the fever is still quite prevalent—it can hardly be eradicated—it can be effectively treated and the prognosis is excellent.

A similar but far milder infection from ticks is found in Africa and the Middle East and is called boutonneuse fever. It is 3 percent fatal when untreated and is carried by any one of several species of ticks.

In 1957 in southern India a tick-borne fever was discovered which was 10 percent fatal and entirely new to the region. The disease is known as the Kyasanur Forest disease. The ticks are thought to be foreign and almost certainly translocated by migratory birds.

Insects

"To the biologist this is not even the age of man, however great his superiority in size and intelligence; it is literally the age of insects."

W. C. Allee

The Surgeon General of the United States Army announced in November 1966 that a new weapon was to be used in the Vietnam War: diamodiphenylysulfone. It turned out to be a very effective weapon against an enemy which was responsible for the deaths of hundreds of G.I.'s and their Vietnamese allies —the disease-carrying mosquito. In a war of kill or be killed, that an insect should have caused so much concern and misery was farcical. It was also intolerable. The only saving grace was that the mosquito was neutral. It was killing the other side too. The drug, diamodiphenylysulfone, which had originally been developed to fight leprosy, cut the fatality rate in Vietnam by 50 percent.

The story illustrates just a facet in man's unceasing and often bizarre war with the insect world—a war that is as old as man. A war that, until this century, was a one-sided war with the insects, at times, having the upper hand. No creature, except man himself, had ever posed such a real threat to the survival of the human race. Even as late as the end of the last century one-half of human deaths could, in one way or another, be blamed upon insects. Flies carried epidemic diseases from the olden-day cesspools and filthy streets. Mosquitoes carried malaria and yellow fever to millions of doomed people. The tsetse spread sleeping sickness, and the flea brought the plague to millions by riding on the back of a rat.

Not until the end of the last century did man begin to

appreciate just how much misery and death insects were causing him. Even now, after some spectacular victories over the insect world, they are still in control of millions of square miles of the earth's surface; even now they kill well over a million people a year. Assuming that until the mid-1800s half the world's deaths were caused by insects—and there is overwhelming evidence that this is so—at least thirty-eight billion people have been killed by these tiny creatures since the Stone Age.

Wars such as the war between man and the insects have good biological precedent. The evolution of life on earth has been profoundly affected by such struggles, which inevitably end in the mass slaughter of one species by another. The irony is that we have the capacity and could quickly develop the equipment to fight a short and very decisive war with the insects if we spent only a fraction of the money and as much time as we are spending upon wars with each other.

The insect, because of its attacks upon man and his livestock, prevents us from developing at least four million square miles of Africa and further millions of square miles in the Amazon and Orinoco areas, where, says W. C. Allee, insects "are undoubtedly in control."

If ever there was an unquestionable case of the female of the species being more deadly than the male it is the mosquito. Only the female bites, for only she requires a blood diet. The male has neither the inclination nor the apparatus and, depending on the species, spends most of his blameless life drinking plant juices.

Bernal R. Weimer estimated in 1951 that one-half of human deaths, exclusive of war and accidents, were caused directly or indirectly by the malaria-carrying mosquito. It is difficult to prove such a statement, but it would certainly have been no exaggeration for the century before, when malaria was killing people by the thousands in the British Isles, in the Low Countries of Europe, as well as in such states as Connecticut and Michigan.

When the World Health Organization began its worldwide campaign against malaria in 1955 it reported: "The truth is that malaria is a disaster, and its terrible consequences (among that third of the world's population which already has more than its fair share of disease and poverty) cannot be easily visualized in countries that are free of it."

Malaria is probably older than man and there is evidence that the dawn men suffered from it. The ancient Indians called it the king of diseases and the ancient Egyptians were plagued by it. It is odd then that a disease so ancient remained such a complete mystery until within living memory. With the notable exception of the Roman farmer-soldier Columella, nobody ever guessed that the mosquito was responsible for malaria until quite recent times. Columella, writing in the first century A.D., said: "There should be no marshland near the buildings . . . for [it] throws off a baneful stench in hot weather and breeds insects armed with annoying stings which attack us in dense swarms . . . from which are often contracted mysterious diseases whose causes are even beyond the understanding of physicians."

But the cue was missed and man entered the medieval period firmly believing that malaria was caused by marsh gases and thus it was called *mal* (bad) *aria* (air). It was 1880 before a French scientist discovered the disease was caused by a parasite. The problem now remained to find out how the parasite got into people. In 1897 Sir Patrick Manson, the British parasitologist, theorized that the mosquito *Anopheles* was the carrier of malaria to man, and a year later Sir Ronald Ross, a British bacteriologist, confirmed that the micro-organisms were in fact carried by mosquitoes; for this he won the Nobel Prize. In 1898 the Italian Giovanni Battista Grassi proved that *Anopheles* transmitted the disease to man and described the cycle for the first time.

Anopheles, an insect which has killed more men than all other man-killing animals combined, looks like any other of the two thousand species of mosquitoes, superficially. It belongs to a family that is spread over the globe and that bites just as fiercely in the Arctic as it does in Vietnam. The actual malaria vectors, though, are confined to the warmer regions and are found in 140 countries.

How does the mosquito infect man? When the female probes a man's flesh to feed from his blood the malarial spores in the insect's saliva pass into the human blood cells, where they multiply rapidly, burst out, and crowd into the blood stream. Along comes a "clean" mosquito for a feed of blood and she will absorb some of the parasites. They then reproduce sexually in her stomach before passing into the salivary gland, where they wait to be injected into their secondary hosts, man. Only the anopheline mosquito can infect man with malaria. Although entomologists know the insect intimately, although they know precisely how it breeds in puddles and slow streams, and although they know just when it feeds at dusk and dawn, they are at a tremendous disadvantage because of the sheer weight of numbers.

The symptoms of malaria were described by WHO in the following way: "First there is the shivering; the chattering teeth, convulsive fits, a skin icy to the touch and a temperature of 104 or higher. After the shivering comes the burning dry heat that drives you almost insane, the insatiable thirst, the booming in the head, the delirium and, worst of all, the hot prickly fire on the skin that is like an excursion into hell." Occasionally the fever can develop into the little-understood complication called blackwater fever, which carries a 50 percent mortality in Africa. In West Africa one in nine hundred malaria cases develops blackwater fever, so called because blood cells discolor the urine.

Malaria until recently was not really considered a sickness; it was looked upon by those in malarial countries as a normal sort of condition—rather as westerners look upon the common cold. But its effects upon the evolution of the races of man have been profound. In a sense the malarial mosquito has helped keep the races of man "pure" in that it made Africa, for instance, a "white man's grave" and thwarted attempts by the genetically unprotected Europeans to colonize the continent on a large scale. In the same way it protected the people of Indonesia against Chinese invaders. But while it has protected some races it has also sapped their strength and kept them in a state of chronic sickness and this discouraged any urge in them to spread their sphere of influence or even to improve their standard of living. How different might history have been if the British Isles had been constantly subjected to endemic malaria. It was malaria that eventually undermined the Roman Empire, that killed Alexander the Great at the age of thirty-three, and that probably killed Oliver Cromwell.

The one hundred million people actually suffering from malaria today are confined mainly to Africa and Southeast Asia. The disease is also found among islanders well out in the Pacific, where it was probably introduced by white explorers. Some authorities claim the disease was unknown in the Americas until the white men came, but this is highly debatable, for the South American Indians had a specific for it in the sixteenth century: they used an extract from the bark of the cinchona tree. This same bark, when it was brought to Europe by the Spaniards, caused a sensation because of its antimalarial properties. Even now many people use it, under the name of quinine.

If ever there was cause for elation about the prospects of defeating malaria it was in the story of Ceylon's fight against the disease. In 1934 the incidence of malaria had died out of most areas in Ceylon. Toward the end of that year, however, a drought dried up the normally fast rivers and many stagnant pools were formed. From these new breeding places came a mosquito population explosion. Four months later a malaria epidemic was reported along the Mata-Oya, and the malaria death rate soared by 800 percent. Out of every 1000 malaria cases reported in 1935 in Kurunegala, 800 died. Within six months of the first signs of the outbreak 4,000,000 people were hit and 60,000 were dead. There were scenes reminiscent of the plague, and hardly a family was left intact. In 1940 an estimated 3,500,000 Ceylonese had malaria—out of a total population of 6,000,000. By 1946 the annual death rate was still in the thousands.

It was then that Paul Muller, the Swiss Nobel laureate, anounced his discovery of dichloro-diphenyl-trichloro-ethane—

DDT. This insecticide of chlorine, benzine, and ethyl alcohol was the "hydrogen bomb" in man's war against the insect world. Supplies were rushed to Ceylon. By 1947, the number of cases was down to 1,500,000. By 1950 there were only 600,000; in 1955, 23,000. In 1960 a mere 467 cases of malaria were reported—out of a total population that had then reached 10.5 million. But in 1968, because of poor maintenance, malaria swept back and thousands now have it again.

When the Eighth World Health Assembly met in Mexico City in 1955, the facts of Ceylon's struggle were among the data studied. It was decided that the World Health Organization "should take the initiative in a program having as its ultimate objective the worldwide eradication of malaria." Total war on the mosquito had been declared. It was no longer a case of controlling malaria; it was now a question of wiping it out.

The plan was simple enough in principle: a preparatory phase would entail surveying the region, briefing staff, and developing resources needed for the coming battle; an attack phase entailed a systematic house-to-house spraying operation which might last up to four years; a consolidation phase would start once the transmission of the disease had been interrupted and entailed treating people still affected by malaria and watching the situation as closely as possible; and then during the maintenance phase a careful check would be instituted to stop the reimportation of the disease. "This phase," says WHO, "comes to an end in all countries when worldwide eradication is achieved." According to Dr. Emilio Pampana, the first director of malaria eradication in WHO, malaria could be eradicated in all countries outside Africa within the next two decades. The total cost might be in the region of one and a half billion dollars. It would cost another six hundred million to eradicate in Africa below the Sahara. Will the world ever be able to afford such a costly war against so small a foe? It would not be unprecedented: we spent more than a billion dollars in killing 22,060,000 people in the last world war.

On the bright side, in no other field of man's endeavor has there been such international cooperation. The malaria picture is changing rapidly and quite dramatically. DDT, however, was not enough. In the 1950's there were still some disturbingly disastrous epidemics. Ten thousand died in the Philippines in 1957; there were a million cases and 10,000 deaths in Greece in 1958. A year later 24,000 died in Mexico. But at least malaria had retreated from the United States, where, in 1955, only 500 cases were reported (mostly imported). The situation in Europe, in many Asian countries, and in most of the Americas improved in the 1960's. The victories show that something can be done. But in any of these areas the disease could flare up again if precautions are relaxed.

Only in Africa is the picture gloomy. A mere 3 million of 197 million Africans who live under the threat of endemic malaria have been freed. Nor is there much effort being made to eradicate the disease over the rest of Africa. A measure of the seriousness of the situation is the fact that in 1964 between 200,000 and 500,000 African infants died from the disease. There can really be no solution until Africa is politically stable. In the long transition periods between governments the mosquito often regains all it has lost. To a certain extent the same applies to Southeast Asia, which is the second biggest problem area as far as WHO is concerned.

There is one more big problem: in 1965 and 1966 an emergence of strains of malarial parasites that were resistant to the world's number one antimalaria drug, chloroquine, were noticed. Resistance to other drugs had long been recognized— especially in Southeast Asia and South America. Although there is no sign of quinine's losing its effectiveness, it was never an entirely "safe" drug. Nor was it capable of killing all parasites in the body. A new drug is urgently being sought, because without one the entire WHO world eradication program could be jeopardized. However, this counterattack by the insect world might well be stopped by a new technique: sterilizing the male insects by radiation and then releasing them in an attempt to reduce the number of descendants.

Anopheles has not been alone in its war with man. It has been assisted for countless centuries by an equally tiny mosquito known as *Aedes aegypti,* the sole carrier of yellow fever. Ironically, although less is known about the cycle of the yellow fever virus than about the cycle of the malarial parasite, man's war against yellow fever has proved highly successful. The disease is in full retreat. Our victory has been accidental in some respects, since the vector tends to disappear when improved sanitation is introduced. The war against *Aedes,* however, has had its stirring moments, and many researchers have calculatedly risked their lives in the battle and lost them.

Yellow fever, unlike malaria, is caused by a virus, which is caused by a relatively sophisticated parasite. This virus is thought to be a degenerate parasite and is passed into the human victim through the infected saliva of the mosquito. Within a few days it may cause an attack of fever so mild that the victim will not know he has a fever. On the other hand it may just as rapidly reach an acute stage. It depends entirely upon how quickly the body can build up the antibodies to fight it. The symptoms are stomach hemorrhage (causing the condition known as "black vomit") and liver damage (usually the fatal factor) that causes the victim to turn yellow with jaundice. At the end of two weeks the patient is either dead or completely cured as well as immune for life.

Yellow fever is not purely a disease of the tropics. It has, over the centuries, claimed tens of millions of lives in devastating epidemics that have ravaged Britain, France, Italy, Yugoslavia, Greece, and many other temperate lands. The same mosquito has caused fever to scythe through the populations of such towns as Philadelphia and Boston as well as villages throughout Africa, Central America, and South America. As the insect does not exist in India and the Far East, yellow fever is unknown there.

During the eighteenth and nineteenth centuries the disease swept through the United States with a regularity that was alarming. In 1793 it struck thirty-five times. The worst was at the end of that year when a ship from the West Indies arrived at

Philadelphia with a sick man aboard—within weeks twenty-four thousand Philadelphians had yellow fever. Gradually the busy town came to a standstill. People fled to the country and the smell of garlic and camphor hung about the streets as the populace went around clutching to their noses cloths drenched in the stuff. Doctors felt sure that yellow fever was a contagious disease, and people saw no reason to disbelieve them. With dumb resignation the afflicted citizens nailed warning notices to their doors. Soon the death toll stood at five thousand. In Philadelphia that year there were twenty major attacks of yellow fever. There were fifteen in New York. Yet nobody suspected the mosquito. It was not until 1881 that Dr. Carlos Finlay in Havana calculated that it could be only the mosquito. He even guessed the species. A few years later, by diligently draining the mosquito's breeding places or spraying them with oil, Havana eradicated yellow fever. It was a major triumph, for few places in the world had been so badly decimated by yellow fever as the Caribbean region. But it was nineteen years before Finlay's hypothesis concerning *Aedes* was proved correct by Major Walter Reed of the United States Army. Meanwhile the mosquito had brought the building of the Panama Canal to a standstill as thousands of laborers came down with yellow fever.

Once the enemy was recognized, the forces of science moved in. Within a few years yellow fever had been eradicated in the West Indies and in the Panama area. In 1905 a short, sharp burst was recorded in New Orleans; it was the last attack in the United States. In 1916 the Rockefeller Foundation donated a sum of money to intensify research into combatting the disease elsewhere, and this enabled laboratories to be opened in both tropical America and tropical Africa. It did not take long for the team in Africa to discover that monkeys also suffered from a form of the disease. This gave researchers the opening they wanted, for until then they often had to give themselves the disease in order to try out antidotes and two or three scientists had died as a result. Rhesus monkeys were brought in from India and within a short time a vaccine had been developed against yellow fever. Today, thanks to this vaccine, yellow fever is an entirely preventable disease.

In 1932 there was a yellow fever outbreak in Espirito Santo, Brazil. The puzzling factor about this outbreak was that *Aedes aegypti* did not exist in that region. Some time before, Bolivian doctors had warned that a type of yellow fever found in South America was caused by a mosquito other than *Aedes aegypti*. It was then that "jungle yellow fever" was described. This strain of yellow fever is confined to the jungle areas in the big river systems of South America and Africa. Man appears to be only an accidental host, for normally it is monkeys which are infected. The main carrier is a mosquito of the genus *Haemagogus* which bites only by day. One thing is clear, jungle yellow fever is likely to always be with us, for the monkey population in the canopy forests will be a reservoir for the disease for generations to come. Nevertheless the jungle strain can be vaccinated against and there have been no really disastrous epidemics for years. The worst recent outbreak was in 1961, when 3000 people died in Ethiopia. The outbreak was stopped after WHO immunized 1,000,000 people. Four years later in Senegal 1287 came down with it and 187 died, but again WHO stepped in and immunized 500,000 Africans, thus halting the epidemic. The situation in South America is slightly better: in 1964 WHO recorded 112 cases and only 89 deaths from Venezuela south to Bolivia.

Apart from dengue or breakbone fever—a rarely fatal, bone-aching disease—the mosquito is responsible for at least one other dangerous disease: filariasis. Filariasis can develop into the dreadfully disfiguring disease of elephantiasis in man and is occasionally fatal. In 1877 Manson traced the disease to the mosquito *Culex,* which he correctly guessed passed the micro-organisms into man when it fed off his blood. The microbes would then produce a larger parasite in the shape of a worm which lies in a lymphatic gland and periodically releases microbes into the bloodstream. The mosquito vectors include both *Aedes* and *Anopheles,* but even today the cycle is not at all clear. In chronic cases the lymph channels become blocked, causing arms, legs, breast, or scrotum to become permanently swollen to tremendous proportions.

There is a growing suspicion in the minds of researchers that some cancers may be caused by a virus or a parasite. If this is correct, a likely carrier is the mosquito. In the East African Virus Research Institute at Entebbe, Uganda, research men are working on a type of cancer called Burkitt's lymphoma, which is one of the commonest diseases affecting Uganda children. Mosquitoes are believed to be the vectors of the disease, which is confined to the tropics.

It has been said that the missionaries who traversed Africa's unexplored regions last century did not just carry the Word of God to the Africans—they carried the scourge of sleeping sickness. While they may have helped the disease break out of its original area in West Africa and the Congo, more than likely it was H. M. Stanley who did the most damage. Toward the end of the 1880s Stanley led an expedition from the Atlantic side of the continent to the Indian Ocean, passing by the southern end of Lake Victoria. He left in his trail 400 dead Africans—many killed by sleeping sickness. It is probably more than a coincidence that a dozen years later, in 1901, 300,000 people living along the southern edge of Lake Victoria were paralyzed by a dreadful outbreak of sleeping sickness. Entire villages fell into a nightmare period of sickness and overwhelming lethargy, and adults and children alike ate their food in a daze and then, day or night, fell into a restless sleep. Out of 300,000 people, 200,000 died. Until then there had been no history of epidemic sleeping sickness in East Africa. Since then it has remained a smoldering threat throughout the entire African tropics and, from time to time, it has flared up like a bush fire. Only in the west, from Senegal to Cameroon and south down through the Congo, is there any long history of the disease. Along with malaria it can possibly be blamed for keeping these areas in a state of primitiveness.

In 1894 the British Royal Commisison on Sleeping Sickness

began work in Zululand, the southern limit of the disease. The commission found that it was caused by a blood parasite called a trypanosome (hence the other name for the disease—trypanosomiasis) and that the parasite was carried by the tsetse fly (*Glossina spp.*), a diurnal, hairy, robust insect about the size of an ordinary housefly. By 1910 it was realized that there were two types of sleeping sickness—one caused by *Trypanosoma gambiense* and the other by *Trypanosoma rhodesiense*. The former may take years to kill its victim; the latter is more rapid in its progress and death usually comes within three months. The fly also causes a form of the disease in cattle which is called nagana. The link between the cattle disease and human disease is imperfectly understood.

Today, because of unstable conditions in Africa, the fifty-year war is now swinging back in the fly's favor. The tsetse fly is counterattacking along several fronts and is in control of at least four million square miles of what could be good cattle country in tropical Africa. According to WHO, the cattle population of this region is about 114 million. If it were not for the presence of the fly, the cattle population could be doubled, which would end the chronic malnutrition of West and Central Africa.

In 1966 WHO stated: "In spite of demonstrable progress, sleeping sickness still is a threat to Africa—a threat that has taken an added significance with the withdrawal of the colonial and trustee powers." WHO appealed for international cooperation on the grounds that until the emergent states had put their health services into gear and had collected the necessary funds, only international help could prevent the disease from getting out of hand.

Here again is the classic problem of how to break the disease's cycle, but in the case of the tsetse fly it is doubly difficult, for the cycle is only incompletely understood. The disease's cycle is more complicated than, for instance, the malarial cycle. There is strong evidence that cattle and perhaps game not only help translocate the fly from area to area but also act as a reservoir. Trypanosomiasis in cattle lacks the symptom of sleepiness and is caused by a different parasite than the one found in human blood but somewhere there is a link between the two. The question is, where? The cattle parasite is apparently not fatal to man. A research worker injected himself with enough nagana-carrying material to kill a herd: he suffered no ill effects.

The fly can drink only blood and tissue juices and in feeding from a human takes into its stomach trypanosomes from the bloodstream. These reproduce inside the fly over a period of fifteen days and produce the micro-organisms in an infective stage. They then pass into the next victim through the fly's saliva. The fly itself is an odd creature which can survive only in shade and lives, on the average, one hundred days. The female (both sexes bite, by the way) bears a single young every ten days; therefore each female can bring into the world only about ten young—in larva form—and these perish unless they can burrow into soil in a few minutes. The larva matures in four weeks, providing the temperature is right and the humidity is right. "Indeed it is a constant wonder to scientists that, despite all these apparent weaknesses, the tsetse has been able to keep itself in existence for thousands of years, and has proved so difficult to eliminate," said a WHO scientist. By trapping the fly and squashing it on filter paper for analysis, researchers have found that *Glossina palpalis* relies upon human blood for only 28 percent of its meals—the rest of the time it feeds off anything from crocodiles, buffaloes, cattle, and wart hogs to porcupines. *Glossina morsitans* feeds off humans 10 percent of the time.

The symptoms of the disease are relatively easy to spot. There is a high fever following inflammation of the lymph glands, and when the parasites eventually attack the spinal cord and the brain there are profound lethargy, melancholy, and angry fits which push the victim to the point of insanity and then death.

It is one thing to control sleeping sickness and quite another thing to eradicate it. The use of drugs and prophylaxis have been effective to a certain extent but the ultimate goal is for drugs to kill the trypanosomes in man so that there is no disease for the tsetse to carry. "But," says WHO, "campaigns based on drug treatment or prophylaxis have not so far proved able to eradicate the disease. Even in those West African countries where they have had the greatest success, sleeping sickness appears to remain obstinately entrenched in a few limited foci." These remaining pockets can, in a very short time, explode into widespread epidemics and thus destroy the work of years and nullify an effort that has cost millions of dollars.

The fight against the tsetse has reached near panic proportions at times and there is still a belief among many scientists that wiping out every head of game in an infected area will solve the problem. Between 1942 and 1950 in Zululand 138,000 head of game were destroyed because they were deemed to be harboring sleeping sickness. Once the game had disappeared, cattle were brought in and the tremendous game-extermination plan had been in vain. In some areas when the fly has been eradicated a wide belt of open land has been created around the area. All shade has been destroyed for several hundred yards, because the fly will not venture across open country— unless he can settle on the underside of animals. This no-man's land is kept clear of all animal life, and even elephants are shot down if they try to cross it.

There are other methods too: insecticides naturally have played a big part and will continue to do so unless present experiments to sterilize hordes of male flies and set them free prove successful.

A disease of which few have heard, and yet seven million have, in Chagas' disease. In Latin America it kills more than one in ten of its victims. The disease is caused by a trypanosome which is excreted by the "kissing bug" (*Triatoma braziliensis*) so called because it bites the soft parts of the face. Chagas' disease is probably impossible to wipe out, for like African sleeping sickness, it has an indestructible reservoir in large and small animals which do not appear to suffer any ill effects.

In man the unicellular parasite attacks the organs and finally the heart. More than thirty-five million Latin Americans are exposed to the disease.

The minute sand fly, whose persistent and quite painful bite makes some of the world's otherwise attractive beaches intolerable, is well known in the western world for a fever it carries. This mild fever is, as far as I know, never fatal. Nevertheless there is a genus of sand fly, *Phlebotomus,* which is a killer of man.

Various species of *Phlebotomus* sandflies infect man with a disease known as Leishmaniasis, a disease which stems from micro-organisms of the genus *Leishmania,* a protozoan of the family Trypanosomidae. The parasites are introduced into their human hosts by the sandfly's bloodsucking activities. The disease is also known as kala azar, and its chief foci are in India, North China, East Pakistan, parts of Middle Asia, the entire Mediterranean seaboard, the Middle East, East and North Africa, Venezuela, Paraguay, and Brazil. Only in India, Russian Turkistan, and the Sudan does kala azar appear to produce epidemics, and unless its victims are treated the results are fatal in 95 percent of cases. It takes up to two years to kill its victims, who suffer such classic symptoms as a grossly enlarged liver and spleen.

The disease is an occupational hazard among forest workers in tropical South America and among the caravaneers of the steppes of Asia. Attempts to eradicate it have not been successful, for domestic dogs, wild dogs, jackals, squirrels, and other rodents can carry the microbes and therefore will forever be a reservoir. Fortunately treatment in human cases is generally effective.

The flea is the vector of the plague, the rat harbors the flea, and man harbors the rat. The three of them are equally responsible for the spread of what is perhaps the most terrifying pestilence known to man. It ravaged the Middle East and North Africa for centuries before it invaded Europe in the sixth century A.D. Then, as man and the rat began to conquer the world together and probe its darkest corners, the plague began to decimate the human race.

The catastrophic pandemic in the mid-1300s was probably triggered off by man's first clumsy attempt at germ warfare in 1343. The Tartars had besieged a large band of Genoese traders in the fortified post of Caffa in the Crimea when, insidiously at first, bubonic plague crept into the Tartar ranks. The Tartars, believing the plague was contagious, placed their dead in catapults and hurled them over the walls of Caffa into the midst of the Genoese. Even that failed to dislodge the traders, but the plague among the Tartars' own ranks forced them to withdraw. Only then did the Genoese—by now riddled with the disease—sail home to their various ports. Some died on the way but enough of them survived to carry the scourge home. The Black Death smoldered at first; then, in 1347, it exploded across Europe and swept through Asia. It was the most costly pandemic, in terms of human life, in the history of man. By 1351 it had killed an estimated 75 million people —25 million of them in Europe. It probably killed off a third of the world's population. In Britain, where the plague came late and left early, 800,000 died. Spontaneously and characteristically the plague inexplicably subsided until it was once more quiescent. Then, three centuries later, it erupted again: in 1664 the Great Plague came to London. On November 2 of that year, a few isolated cases were reported in Westminster. As winter progressed and rats and men huddled in their homes to keep warm, the toll mounted. Alarm became panic and the city authorities, not knowing how to fight the plague, ordered infected houses placed under guard until all the inhabitants were either dead or pronounced healthy. Red crosses were painted on door after door and beneath them were the words "Lord have mercy upon us." In 1665 thousands of Londoners were dying, and the "dead cart" was kept busy taking corpses to mass graves. Thousands more fled the city and thus the plague spread to the country. When summer came the plague died down a little, but as soon as the weather cooled in late autumn the dead cart again went on its rounds accompanied by a bell ringer shouting "Bring out your dead!" By 1666, 74,500 citizens had died—almost one in six out of the total population of 460,000. The belief that the Fire of London in that year burned the plague out is probably unfounded, for the visitation had died out in many parts of Britain of its own accord months before the Fire. But the plague remained on the Continent for years after. In 1675, 1000 died in Malta while thousands more died in Austria, Hungary, Germany, and Poland. In 1679, 76,000 died in Vienna and three years later it killed half of Halle's population of 10,000 and 83,000 in Prague. Year after year it struck towns and villages throughout Europe until in 1720 a crippling outbreak hit Marseilles. Forty thousand died in the port and 10,000 in the surrounding countryside. The plague then disappeared from Europe until, by the twentieth century, it was unknown except in half a dozen places. Parasitologists believe it declined with the black rat, *Rattus rattus,* which was driven out of Europe by the tougher brown rat, *Rattus norvegicus.* The brown rat's fleas are less efficient at transmitting the plague than the black rat's fleas, although both carry the same microbe.

But the respite that Europe enjoyed in the last century was not evident in Asia: Canton lost 100,000 in 1894. In countries such as India, plague will be a serious menace for at least decades to come. In India between 1896 and 1917 nearly 10,000,000 died. In 1923 250,000 died. Even in 1954 there was still evidence of it, and 547 cases were reported. Today, although plague has been reduced to not more than 3000 cases a year in the world (211 deaths in 1966), it is dormant rather than extinct. One thing is certain: should a really serious plague epidemic flare up somewhere in the world, precautionary measures are so efficient today that it could fairly easily be confined.

Few people realize just how near the surface plague is even in countries where it was unknown a century ago. More than seventy species of animals harbor the plague flea, and in the American countryside, rabbit burrows have been found crawling with infected fleas. Thus there is a permanent reservoir of plague even in countries which have never known a serious outbreak. Plague was first recorded in the United States in 1900 when an infected rat came ashore from a foreign ship in San Francisco harbor. Out of the 121 people who contracted the plague that year, 113 died. In 1904 the scare died down and San Francisco was declared plague free—in May 1907 it struck again. By November 1908, 159 cases were treated of which 77 proved fatal. A few deaths were also recorded in New Orleans, Florida, Texas, Oakland, and Los Angeles, where it killed 30 in 1924. It struck Hawaii in 1899, but in forty years only 419 cases were reported, most of them fatal. In South America it was also noted only at the turn of the century and Peru, the country worst affected, had 22,254 cases between 1903 and 1951. Brazil and Argentina have also recorded several thousand.

The bacterium which causes plague, *Pasteurella pestis,* was discovered independently in 1894 by Shibasaburo Kitasato and Alexandre Yersin, who found that it was passed on to a human who was bitten by any one of several species of flea. Since then researchers have realized that the "rural plague" has hardly ever killed a man. The danger lies in the domestic rat picking up the fleas from wild animals. This might cause plague among the urban rat population, and as the rats died the fleas would be forced to look for new hosts. In this way man is infected. Primarily it is a rodent disease and man is only accidentally infected.

There are three types of plague: bubonic, pneumonic, and septicemic. In the latter, death is almost inevitable and is usually very rapid for the bloodstream is heavily invaded by the bacillus. In pneumonic plague the lungs are involved and the disease then becomes extremely contagious—in fact the most contagious disease known to man. In the Manchuria epidemic between 1910 and 1911, pneumonic plague was 99.9 fatal and killed 60,000. Bubonic plague is the most common form and is 20 to 25 percent fatal if untreated. Its symptoms are internal bleeding, delirium, headache, swollen lymph glands and excruciating swellings (buboes) and boils in the groin, armpits, and neck. Recovery is very slow but it leaves the patient immune to further attacks.

Fleas are also responsible for a particularly virulent form of typhus known as murine typhus, which is 50 percent fatal if untreated.

Typhus is another disease spread by a tiny animal—the louse, *Pediculus humanus*—which can count its victims in millions. In fact, it has the doubtful honor of having killed more soldiers in the French trenches in the 1914 to 1918 war than were killed by bullets.

Typhus was endemic throughout medieval times and in the sixteenth century ravaged Spain, from which Cortez is supposed to have sailed with it to the New World. In the eighteenth and nineteenth centuries there were some devastating epidemics throughout Europe and in Britain which were triggered off by conditions in prisons and in the slum areas. Typhus swept Ireland in the middle of the last century after the potato crop failed. When, in 1847, the stricken Irish migrated in large numbers, they took the disease with them. Seventy-five thousand went to Canada and of these 5300 died at sea, 8000 died in Quebec and 7000 in Montreal—in all, more than 25 percent of those who left Ireland died of typhus. But for some reason the disease failed to establish itself in North America, and equally mysteriously, it seems to have disappeared from western Europe in the late nineteenth century. Eastern Europe was not so fortunate and between 1919 and 1923 Poland, Russia, and Rumania lost hundreds of thousands and perhaps even millions because of typhus fever. It returned with a vengeance in the 1930s and 1940s to Germany in the Nazi concentration camps, where it killed tens of thousands. Today, apart from an occasional outbreak in the Far East, Central Europe, Central America, and Africa (10,000 cases were reported in 1964—90 percent from Africa), typhus is under control.

Typhus is caused by a micro-organism called *Rickettsia prowazekii,* which is deposited with the feces of the louse on the human skin and is then scratched in. Within fourteen days the victim is either well again or dead. The prognosis is more favorable the younger one is and, taken for all ages, it is between

5 and 25 percent fatal. Death normally follows a heart attack after days of delirium and coma. Often the first sign of the disease is spots which appear on about the fifth day. Aureomycin has proved the most efficient treatment.

The louse, *Pediculus humanus,* also carries relapsing fever, which is between 6 percent and 30 percent fatal, depending on the condition and the age of the victim. The disease is caused by a parasite which induces a fever that disappears only to reappear a day or so later. There are rarely more than one or two relapses. The parasite can enter the body if the louse is crushed and then scratched into the body.

On January 3, 1964, a group of schoolboys camping near Alberton in the Transvaal disturbed a swarm of bees, which immediately attacked them. One boy in particular was singled out by the bees and enveloped in a seething, moaning cloud. The boys beat the bees off and carried their now unconscious friend to a nearby house. Seven hundred stings were removed from the youngster, who survived the ordeal and was, within a day or two, as right as rain. On January 18, 1966, Mr. A. Barnard, former mayor of Kempton Park in the Transvaal, was stung by a single bee. He lost consciousness and within two days was dead. The same year Mrs. Constance Neil of Randfontain in the Transvaal was watching her brother collecting honey from his hives when she was stung by two bees. She said nothing, for she had often been stung before. A few minutes later, in the car going home, she slumped against the dashboard. She was dead.

Why is it that some people die from a single sting and yet others survive a thousand? Why is it that some people become apparently immune while others become fatally sensitive? There are people who are so sensitive to bee stings that they wear around their necks a locket containing drugs in case they are stung. Even then they have to act fast. In East London in South Africa a twenty-six-year-old farmer, John Mason, carried a loaded syringe wherever he went because of his sensitivity to bee stings. He was plowing one summer day in 1967 when a bee stung him. He lost consciousness before he could inject himself and died soon after.

It is ironic that nowadays the "honey bee" kills more people in the world than the plague-carrying flea. In the United States in 1959 the Office of Vital Statistics gave the annual toll of bee victims as thirty-three. In Africa the annual toll cannot be fewer than a hundred a year. The world total may run as high as a thousand.

The African strain of bee—*Apis adonsonii* and *Apis unicolor* —are notoriously irascible and attempts are being made in South Africa to dilute the strain with the more amiable *Apis mellifica* from Italy. The latter is so docile that in Italy one can watch a class of school children crowding around a beehive while the beekeeper removes the comb. In Africa and in one or two other regions of the world this would be hazardous indeed. The gentlest bees are supposed to come from the Caucasus. Entomologists are at a loss to explain why one species should be ultra-aggressive while another is entirely predictable. The most reasonable explanation so far has been that in Africa, after centuries of nest robbing by man and beast, only the most aggressive strains have been able to survive. In Europe, where beekeeping as opposed to nest-robbing has been practiced since ancient times and an artifical environment has been created for the bees, the insect is very docile.

There is not only doubt about what makes certain bees killers, there is just as much doubt about what exactly it is in the bee's sting that kills. The sting comprises an alkaline fluid and an acidic fluid manufactured by two separate glands. Tests on animals have shown no reaction to injections of either fluid if they are injected separately but the reaction when both are administered at the same time is usually a violent reaction if not death. The European bee-eater, an attractive green bird which relies on bees for a great deal of its diet, may help man find the answer to providing an antiserum for it can be stung on the inside of its throat without reaction yet if it is stung externally it dies within seconds. Attempts to find a really fast and efficient antiserum have so far failed, although antihistamine works in some cases.

Tests have shown that of one hundred patients stung by African strains of bees, 40 percent showed clear signs of hypersensitivity, which is usually associated with respiratory distress and a swelling of the throat. One authority puts the incidence of hypersensitivity at one in one thousand, but he is probably referring only to bees found in Europe. It can hardly be true of Africa. In 1967, for example, three out of eleven Tanzanian school children died within hours after all of them had been stung by angry bees. Not long before, a similar incident occurred farther south. Two girls died in the back of a truck after bees attacked them.

A very useful tip on what to do if chased by a swarm was once given to me by a farmer and was later tried out with gratifying results by the wildlife film producer Sven Persson. The farmer claimed that if one ran for a car, dived in, and wound up the windows the bees would soon become preoccupied with flying against the glass, where they can be swatted. Persson, while filming *The Naked Prey,* was attacked by a swarm and did just this. He became extremely ill from the few stings he had received and might well have received a fatal dose of toxin had he not escaped in this way.

Some famous big game hunter authors have said some silly things about Africa's driver ant. There is hardly a schoolboy in the world who does not believe that driver ants chase elephants, lions; and buffaloes up and down the length of Africa and that they eat men alive in their beds. The outside world looks upon the driver ant as a sort of miniature, crawling piranha. Yet even as I am writing these words I am watching a driver ant beating himself to death against the lamp. In this winged, male version the insect is no more dangerous than a daddy longlegs, and even the blind workers—the ones that march in thick columns and are supposed to terrify big game —are more fascinating than frightening.

One famous writer-hunter who stumbled across a column

of them marching in search of food one night wrote how glad he was that he had his gunbearers with him: "Otherwise I might have been devoured alive." The same man writes: "By and large I would rather face a charging lion than wilfully disturb a column of these ants on the march." In another passage: "Nothing avails save instant flight. It is no time for heroics. Man's sole effective defence . . . is a resort to fire, water, petrol or poison gas."

Although it seems logical that the siafu or driver ant (subfamily Dorylinae) must have, at some time or another, killed African babies or even finished off and eaten dying men, I have found no such incident on record. The ants are nomadic and live temporarily under rocks or trees from where they periodically emerge in massive numbers to go in search of food. They march in a column that can stretch thousands of yards and which is six to eight ants in width and flanked by "guards." They will devour anything they can overwhelm by sheer weight of numbers and have been known to eat dogs and, on one occasion, a gorged and helpless python. It is possible that they choke and blind their prey first by crawling into its nostrils, throat, and eyes.

Loveridge gives a magnificent account of how his house in East Africa was invaded by driver ants. He had to run up and down on the spot when he stood on the floor to stop them crawling up his legs and he saw beetles and roaches falling from every nook and cranny covered in biting siafus. He could hear the rats squeaking in terror in the roof as they were attacked. Finally, after a very uncomfortable night, he was forced to sleep outside until the ants had cleansed the house of vermin. Later he found they had killed (but not eaten) a young pet crocodile in the garden.

There is only one sure way to stop a column of driver ants and that is by soaking the ground ahead of them with petrol and setting it alight. If the fire is allowed to die down the column is quite likely to go pouring into it like a runnel of treacle until the fire is extinguished.

Kirkpatrick found that driver ants were totally different from army ants in the Americas in that where army ants give one a nip as soon as they find something worth nipping, African driver ants crawl up one's legs without being felt. When they get about waist high—by which time one's lower regions are crawling with them—they all begin biting in unison. Elephants will often fly into a rage when the ants crawl up their trunks, and this has probably helped foster the theory that when the driver ant marches all creatures in its path flee for their lives.

The only ant in the New World that can be considered dangerous is the fire ant (subfamily Myrmicinae), which has a painful sting and is suspected of having caused the death of a man. The United States Department of Agriculture seems unduly worried by the ant and has gone to enormous lengths to try to eradicate it, including the treatment of millions of acres with insecticides.

X
THE PREY

War Graves in Normandy

The Prey

"The human race's prospects of survival
were considerably better when we were
defenceless against tigers than they are
today when we have become defenceless
against ourselves."

Arnold Toynbee

The world's most habitual man-killer among mammals is unique
in the animal kingdom. Its large brain and intelligence are
unrivaled; its ability to live under almost any conditions, at any
height or latitude, is unmatched, and its curious preoccupation
with the killing of its own species has no parallel in biological
history. That man-killer is man.

Bearing in mind our close affinity to the animal world, our
constant wars can be partly explained. Like all animals we are
basically interested in territory in one form or another and
when we feel our particular territory is being threatened or
that we require or are entitled to a larger territory, we will
fight for that territoy, if necessary to the death. But outside the
heat of war man's record for man-killing is not so easily
explained.

What mass psychosis for instance drove the Aztecs to tie
down eighty thousand human beings and in one ceremony cut
out their hearts? What could have driven these highly intelligent
and civilized people to sacrifice annually two hundred thousand
of their own kind under the most horrific circumstances? Wild
killing orgies of frightening magnitude are found liberally
sprinkled through the histories of Asia, Africa and Oceania,
America and Europe. Many of them must have stemmed from
the same motive as the spectacles of mass slaughter in the
Roman arena when the Romans used animals to do their human

killings for them—spectacles that were apparently calculated to condition a warlike people and nourish the martial spirit. Some of the more primitive mass-killing orgies were followed by cannibalism and may well have sprung from an unconscious craving for protein.

Man as a man-eater, man as a cannibal, is a subject that has received only niggardly attention from anthropologists. Cannibalism is still practiced today in parts of the Far East and in Africa and cannot be attributed to any single reason. Man is said to have been cannibalistic from the beginning: the Australopithecines are thought to have eaten each other. The Cro-Magnon man, who lived from 60,000 years ago until about 4000 B.C., probably ate the last of the Neanderthal-type man, whom he succeeded. Man is now, however, an habitual cannibal. Throughout man's evolution the incidence of cannibalism has been a recurring phenomenon rather than a tradition.

Most Africans today abhor the thought of cannibalism although the grandfathers of many of them practiced it. A notable period for cannibalism in Africa was the last century following the depredations of such "generals" as Mzilikatze in the Transvaal, who, with his warrior hordes, broke up the peaceful pastoral tribes, causing them to drift leaderless and starved over the veld preying upon even weaker refugee units and, on occasion, eating them. The only cannibalism recorded recently in southern Africa has been in isolated and very rare ritualistic ceremonies. Higher in the continent, notably in the Congo, cannibalism is still prevalent in some remote areas. During the Mau Mau disturbances, thousands of initiates at the instigation of witch doctors ate parts of corpses and drank fresh blood.

In Oceania, where some island communities frequently cook and eat humans, the custom seems to stem from necessity. In the Melanesian region some tribes still regard human meat as a natural source of protein and use the term "long pig" to describe an edible human corpse, which is usually the body of an enemy tribesman who was hunted down for the express purpose of eating him.

In Papua thieves were punished right up until recent times

by having to submit their women for slaughter by the aggrieved parties, who then ate them. In the same region it is the custom among some conservative tribesmen to eat their dead relatives in order to show their respect and to ingest some of the good abstract qualities within them as well as some of the nutritional ones. A similar custom is occasionally practiced by the Australian aborigines; here again circumstances suggest the custom might have nutritional advantages for an otherwise protein-hungry people. The Maoris of New Zealand, a highly civilized people in most respects, ate their enemies at ceremonial feasts up until living memory. Their land was notably lacking in animals which could be killed for protein. Hans Staden in *The True History of His Captivity* (1557) describes how South American Indians— the Tupinamba of Brazil—looked after certain captives and allowed them to marry even though they were earmarked for eating. Although these captives knew their fate, they made no attempt to escape and played the game when their final day arrived. On that day they were ceremoniously clubbed to death, the bodies prepared and cooked, and friends and relatives invited to the feast.

The taste of human flesh, according to cannibals, is more subtle than game. George Rushby quotes a Tanzanian cannibal as saying: "There are streaks of fat running through the lean meat. It tastes better and is much sweeter than elephant meat or any other meat, including gorilla and chimpanzee. Also, as you know, when you place a piece of elephant meat in the pot to boil, it just lies there at the bottom like a stone while it is being cooked—it is quite dead. But if you boil a piece of human meat it bumps and bounces around and jigs about in the pot; it has life."

A reason for our lack of knowledge of cannibalism is that as soon as civilization influences an area, cannibalism, where it exists, is declared illegal and dies out before it can be properly studied.

A subject allied to cannibalism, in some respects, is lycanthropy, which we have touched on in the chapter about lions. This form of hysteria is often present in pregnant women and manifests itself by causing the patient to yearn for flesh—preferably human flesh. Quite often the lycanthropist is convinced he or she is an animal. The mania (as it can be sometimes described) is probably more prevalent in Africa, where there are several instances of groups becoming "lion-men" or "were-hyenas" or "leopard-people." As we have seen, some of those involved are not always as mad as one would suppose. In Europe the most common forms of lycanthropy occur in men believing they are wolves or bears. Jaguar-people are known in South America and tiger-men are known in primitive areas in the Far East.

In the pursuit of its pastime of man-killing the human race has shown some remarkable, if disturbing, originality. It has frequently co-opted lower creatures in the animal world to carry out killings. Pliny, for instance, refers to slaves being fed to specially starved, needle-toothed moray eels and Julius Caesar is said to have fed six hundred slaves to them as a form of entertainment. We have also seen numerous examples of man's use of man-killers in war.

Although history is punctuated by some bleak periods of human depravity it is difficult to find a period when compassion was so dead as in the years of the Roman Empire, when, for the entertainment of the masses, men and women were fed to wild animals in the arena.

The era began sometime after 300 B.C., when criminals, prisoners of war, and young adventurers were formed into a corps of gladiators. A century or so later, a corps of *bestiarii*—gladiators who were trained to fight man-killing animals in the arena—emerged. Even this began to pall, and soon after a new attraction was added to arena program: executions carried out by all manner of animals. These gruesome events continued for half a thousand years until in A.D. 326, Constantine repealed the law that permitted criminals to be condemned to die by wild animals. Nevertheless, for five hundred years criminals and religious nonconformists were tied to stakes or merely pushed into the arena to be torn apart by lions, tigers, leopards, bears, and any one of half a dozen other species. Occasionally, some victims, you may remember, were trampled or torn apart by trained elephants. But such spectacles created mixed feelings, for there was still just a spark of compassion glowing faintly in the deepest recesses of the dark Roman heart. It led some spectators to protest that the men-versus-elephant events were debasing and an insult to the peculiar dignity of the elephant.

The first animals to die by the sword of the *bestiarii* were probably lions and leopards, and judging by the weapons allowed for such encounters, the men usually stood a slightly better than even chance against the leopard, but very little chance indeed against the lion. By the time of Christ, these events were very popular indeed.

In A.D. 80, Titus opened the Colosseum in Rome. The opening of this enormous stadium (six hundred by five hundred feet, with an arena well over twice the size of an English football field) heralded an era of unparalleled barbarity in man's association with animals. The gala opening was a blood bath: five thousand animals and an unrecorded number of men died on the first day. Unarmed and inadequately armed prisoners were pitted against man-eating beasts, and Christians—draped in animal skins and tied to posts—were eaten alive by lions while fifty thousand spectators howled and cheered in the popular stands. For the next three months there was a daily toll of human and animal life in the arena.

Tigers, lions, leopards, bears, elephants, rhinoceroses, hippopotamuses, snakes, crocodiles, bulls, and stags were introduced at enormous cost in money and in life. They were introduced by various public figures each trying to outdo the other. "The mob yawned at the tedious variety of nature," writes E. S. Turner, but it did not yawn too much. If the popularity of the arena had waned to any great extent, Rome could have been ripped apart. The arena events had become a safety valve for the highly strung, bloodthirsty Roman mob. As the horrors of the arena had less and less effect, the Roman emperors, one after another, presented spectacles that became more and more bizarre. Probus in A.D. 276 planted a forest in the arena of the Colosseum and in it released one hundred lions, two hundred leopards, and three hundred bears so that the audience could watch hunters kill them with spears, or be killed. Gordian in A.D. 237 staged fights involving up to a thousand bears.

In these later battles between men and beasts, as many as eleven thousand wild animals were gathered together in a single program, which might last for days. Occasionally prisoners or even volunteers faced them with a cloak that was used for confusing such animals as lions and bears and, having gotten them off guard, for strangling them. Some men faced wild beasts with only their bare fists for protection. Human life had never been cheaper: at the one thousandth anniversary of Rome, two thousand gladiators died in combat with each other and with wild animals for the glory of the empire.

The idle emperor Commodus was popular in the arena. He frequently showed off his skill in it with bow and crescent-headed arrows. Once he shot dead one hundred lions with as many arrows, a feat which would have been remarkable even if he had been armed with a rifle. On another occasion Commodus had a leopard loosed upon a criminal tied to a stake. He waited for the leopard to leap up at the resigned victim before he dropped it at the man's feet with a heart shot.

Although other countries staged mortal tournaments between men and animals, none came near the savagery of Rome and nowhere did the popularity of such spectacles last. Nowhere that is, except in Spain.

The twentieth-century matador is no less than a bestiarius and the spectacle of bullfighting is a vestige of the Roman amphitheater contests. Spain and to a much lesser extent Portugal, Mozambique, and Mexico are the last places left in the world where the blood of man and beast still frequently merges in the name of entertainment.

The bulls used in the ring are usually a Seville breed noted for their inborn aggressive: an aggressiveness that has long been recognized. Slightly built, sharp-horned, and nimble, the Seville bull had been used with a modicum of success in ancient battles and a phalanx of them put to flight the Carthaginian

Hamilcar Barca when he attacked the town they now call Barcelona. The Celts, armed with axes, fought the same breed, one man against one bull, as a form of sport. But it was the Romans who, in A.D. 41, saw the more artistic possibilities in bullfighting. They took some Spanish bulls back to Rome and trained men to wrestle them barehanded.

The sport found its way back to Spain and barehanded contests remained popular there until 1700, when the sword was used for the first time. Since that year, out of Spain's 125 champion bullfighters, one in three has died in the arena. Since 1900, two hundred or more bullfighters have been killed in public. Juan Belmonte, who was gored to death at the peak of an amazing career, had been horned fifty times.

Few matadors who fight in Spain's four hundred *plazas de toros* survive a season without a serious wound. Bernhard Grzimek, who, perhaps a little sarcastically, refers to such incidents as "industrial accidents," claims their wounds would be worse if the promotors did not saw the bulls' horn tips off before bullfights and then glue them on in such a way that they easily break off. Grzimek also accuses the Spanish of doping bulls before a contest. Whether the numerous accusations leveled at the Spanish have foundation or not, the spectacle of bullfighting has lost none of its aura of cultural respectability. The *plazas de toros* continue to make millionaires out of men such as Manolete, who, in 1947, at the age of thirty, had three million dollars in his bank. He was gored to death that same year.

In the last three generations man has calculatedly and with malice killed at the average rate of at least three-quarters of a million people a year. In the last world war he killed 22,060,000. Six million of these were exterminated in one carefully planned program of genocide. In an earlier war between the world's most civilized nations, in the space of one day, 21,392 men were left dead on the banks of the river Somme. More recently, using his most effective weapon to date, man killed at least 70,000 in a single explosion and left tens of thousands mortally burned.

In individual acts of murder for gain or through passion his annual toll of his own species is in excess of 150,000. It is little wonder that in acts of suicide—acts which, for once, have no biological precedent—man kills himself at the rate of 360,000 a year.

Man's preoccupation with making weapons and his appetite for the kill might well stem from the fact that he was born with a weapon in his hand and owes his survival to his ability in using it. Without the aid of manufactured weapons the terrestrial dawn man, clawless, fangless, and lacking in speed, could never have survived. Unlike his monkey cousins whose long arms allowed them to swing into trees for refuge, man had nowhere to hide: he had to stand and fight those animals which challenged him. Later he grew so efficient that he was able to drive predators from their caves and move in himself. Dawn man's one great talent was the talent to kill in a variety of ways and with a variety of weapons. Finally he was able to subdue most of the animal kingdom.

Instinct, being the product of inherited memory, has produced in nearly all land animals—from the elephant to the lion and from the bear to the taipan—a fear of man. Only the fishes whose world has yet to be invaded by us show little or no fear.

The irony of man's victory over the greater part of the animal world is that he himself has emerged as his own worst enemy.

Bibliography

ALLEE, WARDER C., *Cooperation Among Animals,* Schuman, 1951.

ANDERSON, KENNETH, *Man-eaters and Jungle Killers,* Allen & Unwin, 1957.

———, *This Is the Jungle,* Allen & Unwin, 1964.

BALNEAVES, ELIZABETH, *Elephant Valley,* Rand McNally, 1963.

BARROS PRADO, EDUARDO, *The Lure of the Amazon,* Souvenir Press, 1959.

BLOND, GEORGES, *The Elephants,* Macmillan, 1961.

BOELDEKE, ALFRED, *With Graciela to the Head-hunters,* McKay, 1959.

BRANDER, H. D., *Wild Animals in Central India,* 1923.

BRELAND, OSMOND P., *Animal Facts and Fantasies,* Harper, 1950.

———, *Animal Friend and Foes.*

BRICK, HANS, *Jungle Be Gentle,* Peter Davies, 1960.

BULPIN, T. V., *The Hunter Is Death,* Nelson, 1962.

BURKE, EDMUND H., *The Underwater Handbook,* Wehman, 1963.

BUTLER, JEAN CAMPBELL, *Danger, Shark!,* Little, Brown, 1964.

CALDWELL, HARRY R., *Blue Tiger,* Abingdon, 1924.

CARR, ARCHIE F., *Ulendo,* Knopf, 1964.

CARRIGHAR, SALLY, *Moonlight at Midday,* Knopf, 1958.

———, *Wild Heritage,* Houghton Mifflin, 1965.

CARRINGTON, RICHARD, *Elephants,* Chatto & Windus, 1958.

CARSON, RACHEL, *The Silent Spring,* Houghton Mifflin, 1962.

COPPLESON, VICTOR, *Shark Attack,* Sydney, Angus and Robertson, 1962.

COON, CARLETON S., *The Living Races of Man,* Knopf, 1965.

CORBETT, JAMES E., *My India,* Oxford University Press, 1952.

COUSTEAU, JACQUES-YVES, and DUMAS, FREDERIC, *The Silent World,* Harper, 1953.

DAVIES, DAVID H., *About Sharks and Shark Attack,* Hobbs, 1966.

DEMBECK, HERMAN, *Animals and Men,* Doubleday, 1966.

DENIS, ARMAND, *Cats of the World,* Houghton Mifflin, 1964.

DITMARS, RAYMOND L., *Snakes of the World,* Macmillan, 1931.

DOBIE, J. FRANK, *Rattlesnakes,* Little, Brown, 1965.

FAIRLEY, N. H., "Snakes," in *Australian Medical Journal,* March 9, 16, 23, and June 8, 1929.

FAWCETT, LIEUTENANT COLONEL PERCY HARRISON, *Exploration Fawcett,* Funk & Wagnalls, 1953.

FORAN, W. ROBERT, *A Breath of the Wilds,* Robert Hale, 1959.

———, *A Hunter's Saga,* Robert Hale, 1961.

FROMAN, ROBERT, *The Nerve of Some Animals,* Gollancy, 1961.

GEE, E. P., *The Wild Life of India,* Dutton, 1965.

GRZIMEK, BERNHARD, *Such Agreeable Friends,* Hill & Wang, 1965.

———, *Twenty Animals and One Man.*

———, *Wild Animal, White Man,* Hill & Wang, 1966.

HENSCHEN, FOLKE, *The History and Geography of Diseases,* Dell, 1966.

HILLABY, JOHN D., *Journey to the Jade Sea,* Simon & Schuster, 1965.

HUNTER, JOHN A., and MANNIX, DANIEL P., *African Bush Adventure.*

ILLINGWORTH, FRANK, *Wild Life Beyond the North,* Country Life, 1951.

IONIDES, C. J. P., *A Hunter's Story,* Harper, 1966.

JOHNSON, D., *Indian Field Sports,* London, 1927.

JORDAN, EMIL L., *Nature Atlas of America,* Hammond, 1952.

KIRKPATRICK, THOMAS W., *Insect Life in the Tropics,* Longmans, Green.

LAWMAN, TONY, *From the Hands of the Wicked,* Robert Hale, 1960.

LURIE, RICHARD, *Under the Great Barrier Reef,* International Publications Service, 1966.

LYELL, DENIS D., *African Adventure,* John Murray, 1935.

MACPHERSON, A. H., in *Oryx,* Vol. 8, No. 5, Canadian Fauna Preservation Society.

MANNIX, DANIEL P., *All Creatures Great and Small*, Longmans, Green.

MATTHIESSEN, PETER, *Wildlife in America*, Viking.

MERFIELD, FRED G., *Gorillas Were My Neighbours*, Longmans, Green.

MONTAGU, S. R., *North to Adventure*, 1939.

MOOREHEAD, ALAN, *No Room in the Ark*, Dell, 1968.

MOWAT, FARLEY, *Never Cry Wolf*, Atlantic, Little, Brown, 1963.

NEIMER, BERNAL R., *Man and the Animal World*, Wiley, 1951.

NELSON, E. W., *Wild Animals of North America*, Washington, 1930.

PATTERSON, J. H., *The Man-eaters of Tsavo*, Macmillan, 1907.

PEDERSEN, ALWIN, *Polar Animals*, Taplinger, 1966.

PERRY, RICHARD, *The World of the Polar Bear*, University of Washington Press, 1966.

————, *The World of the Tiger*, Atheneum, 1965.

PETERSON, RUSSEL, *Silently, by Night*, Longmans, Green, 1966.

PHILLIPS, CRAIG, *The Captive Sea*, Chilton, 1964.

POLLARD, JOHN, *Wolves and Were-wolves*, Robert Hale, 1966.

RUARK, ROBERT C., *The Vulgar Assassin*, Pan, 1965.

RUSSELL, P. F., in *Bulletin of the New York Academy of Medicine*, 19:599–630, 1943.

SANDERSON, IVAN T., *The Dynasty of Abu*, Knopf, 1962.

SAVORY, THEODORE H., *Spiders, Men and Scorpions*, University of London, 1961.

SCHALLER, GEORGE, *The Year of the Gorilla*, University of Chicago Press, 1964.

SETON, ERNEST THOMPSON, *Lives of Game Animals*, New York, 1929.

SILVA, JOÃO AUGUSTO, *Gorongoza*, Empresa Moderna, 1965.

SINGH, K., *The Tiger of Rajasthan*, 1959.

SMITH, J. L. B., *The Sea Fishes of Southern Africa*, CNA.

SMITH, T. MURRAY, *The Nature of the Beast*, Coward, 1963.

STEWART, COLONEL A. E., *Tiger and Other Game*, 1927.

STRACEY, P. D., *Elephant Gold*, Weidenfeld & Nicholson, 1963.

SULLIVAN, WALTER, *Quest for a Continent*, McGraw-Hill, 1957.

TAYLOR, JOHN, *Man Eaters and Marauders*, Muller, 1959.

THORPE, RAYMOND W., and WOODERSON, W. D., *The Black Widow*, University of North Carolina Press, 1945.

WEBSTER, DAVID KENYON, *Myth and Man-Eater*, Peter Davis.

WEIMER, BERNAL R., *Man and the Animal World*, Wiley, 1961.

WESSELS, PETE, *The Johannesburg Star*, Supplement 17:11–66.

WHITE, GRANT, *Jungle Down the Street*, Phoenix, 1957.

WILLIAMS, J. H., *Big Charlie*, Hart Davis, 1959.

WILLOCK, COLIN O., *The Enormous Zoo*, Harcourt, 1965.

WOLHUTER, HARRY, *Memories of a Game Ranger*, CNA, 1948.

WOOD, REVEREND J. C., *Wood's Illustrated Natural History*, Routledge, 1854.

WRIGHT, BRUCE S., *Ghost of the North*, 1959.

WYKES, ALAN, *Snake Men*, Simon & Schuster, 1961.

Index